A-Z
NANO RESEARCH

A-Z NANO RESEARCH

Dr. Albert Shawn

CENTRUM PRESS
NEW DELHI-110002 (INDIA)

CENTRUM PRESS

H.O.: 4360/4, Ansari Road, Daryaganj,
New Delhi-110 002 (India)
Ph.: 23278000, 23261597

B.O.: No. 1015, Ist Main Road, BSK IIIrd Stage
IIIrd Phase, IIIrd Block,
Bangalore - 560 085 (India)
Tel.: 080-41723429
Visit us at: www.centrumpress.com

A-Z Nano Research

First Edition, 2009

ISBN 978-93-80106-97-7

PRINTED IN INDIA

Printed at Mehra Offset Press, Delhi

Contents

Preface

"This book has been brought out to acquaint the readers with the hype and the rigmarole behind Nano research. The emphasis of the book is on presenting forth a thorough introductory manual on the subject, detailing its history, the science and the practice, the techniques, the trends and developments, of this science.

The multifold implications of nanoresearch have been discussed most comprehensively, as this science holds the key to understanding the future of humanity in terms of scientific developments. The critical insights presented forth in the book are highly informative, and it serves well for all those interested in understanding Nano reseach in all its machinations, and manifestations."

Author

Chapter 1

Cell Repair Machines

We will use molecular technology to bring health because the human body is made of molecules. The ill, the old, and the injured all suffer from mis-arranged patterns of atoms, whether mis-arranged by invading viruses, passing time, or swerving cars. Devices able to rearrange atoms will be able to set them right. Nanotechnology will bring a fundamental breakthrough in medicine.

Physicians now rely chiefly on surgery and drugs to treat illness. Surgeons have advanced from stitching wounds and amputating limbs to repairing hearts and re-attaching limbs. Using microscopes and fine tools, they join delicate blood vessels and nerves. Yet even the best micro-surgeon cannot cut and stitch finer tissue structures. Modern scalpels and sutures are simply too coarse for repairing capillaries, cells, and molecules.

Consider "delicate" surgery from a cell's perspective: a huge blade sweeps down, chopping blindly past and through the molecular machinery of a crowd of cells, slaughtering thousands. Later, a great obelisk plunges through the divided crowd, dragging a cable as wide as a freight train behind it to rope the crowd together again. From a cell's perspective, even the most delicate surgery, performed with exquisite knives and great skill, is still a butcher job. Only the ability of cells to abandon their dead, regroup, and multiply makes healing possible.

Yet as many paralysed accident victims know too well, not all tissues heal.

Drug therapy, unlike surgery, deals with the finest

structures in cells. Drug molecules are simple molecular devices. Many affect specific molecules in cells. Morphine molecules, for example, bind to certain receptor molecules in brain cells, affecting the neural impulses that signal pain. Insulin, beta blockers, and other drugs fit other receptors. But drug molecules work without direction. Once dumped into the body, they tumble and bump around in solution haphazardly until they bump a target molecule, fit, and stick, affecting its function.

Surgeons can see problems and plan actions, but they wield crude tools; drug molecules affect tissues at the molecular level, but they are too simple to sense, plan, and act. But molecular machines directed by nanocomputers will offer physicians another choice. They will combine sensors, programs, and molecular tools to form systems able to examine and repair the ultimate components of individual cells. They will bring surgical control to the molecular domain.

These advanced molecular devices will be years in arriving, but researchers motivated by medical needs are already studying molecular machines and molecular engineering. The best drugs affect specific molecular machines in specific ways. Penicillin, for example, kills certain bacteria by jamming the nanomachinery they use to build their cell walls, yet it has little effect on human cells.

Biochemists study molecular machines both to learn how to build them and to learn how to wreck them. Around the world (and especially the Third World) a disgusting variety of viruses, bacteria, protozoa, fungi, and worms parasitize human flesh. Like penicillin, safe, effective drugs for these diseases would jam the parasite's molecular machinery while leaving human molecular machinery unharmed. Dr. Seymour Cohen, professor of pharmacological science at SUNY (Stony Brook, New York), argues that biochemists should systematically study the molecular machinery of these parasites. Once biochemists have determined the shape and function of a vital protein machine, they then could often design a molecule shaped to jam it and ruin it. Such drugs could free humanity from such ancient horrors as schistosomiasis and leprosy, and from new ones such as AIDS. Drug companies are already redesigning molecules based on knowledge of how they work. Researchers at Upjohn Company have designed and made modified molecules of vasopressin, a hormone that consists of a short chain of amino acids. Vasopressin increases the work done by the heart and decreases the rate at which the kidneys produce urine; this increases blood pressure. The researchers designed modified vasopressin molecules that affected receptor molecules in the kidney more than those in the heart, giving them more specific and controllable medical effects. More recently, they designed a modified vasopressin molecule that binds to the kidney's receptor molecules without direct effect, thus blocking and *inhibiting* the action of natural vasopressin.

Medical needs will push this work forward, encouraging researchers to take further steps toward protein design and molecular engineering. Medical, military, and economic pressures all push us in the same direction. Even before the assembler breakthrough, molecular technology will bring impressive advances in medicine; trends in biotechnology guarantee it. Still, these advances will generally be piecemeal and hard to predict, each exploiting some detail of biochemistry. Later, when we apply assemblers and technical AI systems to medicine, we will gain broader abilities that are easier to foresee.

To understand these abilities, consider cells and their self-repair mechanisms. In the cells of our body, natural radiation and noxious chemicals split molecules, producing reactive molecular fragments. These can misbond to other molecules in a process called cross-linking. As bullets and blobs of glue would damage a machine, so radiation and reactive fragments damage cells, both breaking molecular machines and gumming them up.

If our cells could not repair themselves, damage would rapidly kill them or make them run amok by damaging their control systems. But evolution has favored organisms with machinery able to do something about this problem. The self-replicating factory system sketched are repaired itself by replacing damaged parts; cells do the same. So long as a cell's DNA remains intact, it can make error-free tapes that direct ribosomes to assemble new protein machines.

Unfortunately for us, DNA itself becomes damaged, resulting in mutations. Repair enzymes compensate somewhat by detecting and repairing certain kinds of damage to DNA. These repairs help cells survive, but existing repair mechanisms are too simple to correct all problems, either in DNA or elsewhere. Errors mount, contributing to the aging and death of cells - and of people.

LIFE, MIND, AND MACHINES

Does it make sense to describe cells as "machinery," whether self-repairing or not? Since we are made of cells, this might seem to reduce human beings to "mere machines," conflicting with a holistic understanding of life.

But a dictionary definition of holism is "the theory that reality is made up of organic or unified wholes that are greater than the simple sum of their parts." This certainly applies to people: one simpler sum of our parts would resemble hamburger, lacking both mind and life.

The human body includes some ten thousand billion *billion* protein parts, and no machine so complex deserves the label - mere." Any brief description of so complex a system cannot avoid being grossly incomplete, yet at the cellular level a description in terms of machinery makes sense. Molecules have simple moving parts, and many act like familiar types of machinery. Cells considered as a whole may seem less mechanical, yet biologists find it useful to describe them in terms of molecular machinery.

Biochemists have unraveled what were once the central mysteries of life, and have begun to fill in the details. They have traced how molecular machines break food molecules into their building blocks and then reassemble these parts to build and renew tissue. Many details of the structure of human cells remain unknown (single cells have billions of large molecules of thousands of different kinds), but biochemists have mapped every part of some viruses. Biochemical laboratories often sport a large wall chart showing how the chief molecular building blocks flow through bacteria. Biochemists understand much of the process of life in detail, and what they don't understand seems to operate on the same principles. The mystery of heredity has become the industry of genetic engineering. Even embryonic development and memory are being explained in terms of changes in biochemistry and cell structure.

In recent decades, the very quality of our remaining ignorance has changed. Once, biologists looked at the process of life and asked, "How can this be?" But today they understand the general principles of life, and when they study a specific living process they commonly ask, "Of the many ways this could be, which has nature chosen?" In many instances their studies have narrowed the competing explanations to a field of one. Certain biological processes - the coordination of cells to form growing embryos, learning brains, and reacting immune systems - still present a real challenge to the imagination. Yet this is not because of some deep mystery about how their parts work, but because of the immense complexity of how their many parts interact to form a whole.

Cells obey the same natural laws that describe the rest of the world. Protein machines in the right molecular environment will work whether they remain in a functioning cell or whether the rest of the cell was ground up and washed away days before. Molecular machines know nothing of "life" and "death."

Biologists - when they bother - sometimes define life as the ability to grow, replicate, and respond to stimuli. But by this standard, a mindless system of replicating factories might qualify as life, while a conscious artificial intelligence modeled on the human brain might not. Are viruses alive, or are they "merely" fancy molecular machines? No experiment can tell, because nature draws no line between living and nonliving. Biologists who work with viruses instead ask about viability: "Will this virus function, if given a chance?" The labels of "life" and "death" in medicine depend on medical capabilities:

physicians ask, "Will this patient function, if we do our best?" Physicians once declared patients dead when the heart stopped; they now declare patients dead when they despair of restoring brain activity. Advances in cardiac medicine changed the definition once; advances in brain medicine will change it again.

Just as some people feel uncomfortable with the idea of machines thinking, so some feel uncomfortable with the idea that machines underlie our own thinking. The word "machine" again seems to conjure up the wrong image, a picture of gross, clanking metal, rather than signals flickering through a shifting weave of neural fibers, through a living tapestry more intricate than the mind it embodies can fully comprehend. The brain's really machine like machines are of molecular size, smaller than the finest fibers.

A whole need not resemble its parts. A solid lump scarcely resembles a dancing fountain, yet a collection of solid, lumpy molecules forms fluid water. In a similar way, billions of molecular machines make up neural fibers and synapses, thousands of fibers and synapses make up a neural cell, billions of neural cells make up the brain, and the brain itself embodies the fluidity of thought.

To say that the mind is "just molecular machines" is like saying that the Mona Lisa is "just dabs of paint." Such statements confuse the parts with the whole, and confuse matter with the pattern it embodies. We are no less human for being made of molecules.

FROM DRUGS TO CELL REPAIR MACHINES

Being made of molecules, and having a human concern for our health, we will apply molecular machines to biomedical technology. Biologists already use antibodies to tag proteins, enzymes to cut and splice DNA, and viral syringes (like the T4 phage) to inject edited DNA into bacteria. In the future, they will use assembler-built nanomachines to probe and modify cells.

With tools like disassemblers, biologists will be able to study cell structures in ultimate, molecular detail. They then will catalog the hundreds of thousands of kinds of molecules in the body and map the structure of the hundreds of kinds of cells. Much as engineers might compile a parts list and make engineering drawings for an automobile, so biologists will describe the parts and structures of healthy tissue. By that time, they will be aided by sophisticated technical AI systems.

Physicians aim to make tissues healthy, but with drugs and surgery they can only encourage tissues to repair themselves. Molecular machines will allow more direct repairs, bringing a new era in medicine.

To repair a car, a mechanic first reaches the faulty assembly, then identifies and removes the bad parts, and finally rebuilds or replaces them. Cell repair will involve the same basic tasks - tasks that living systems already prove possible.

- Access. White blood cells leave the bloodstream and move through tissue, and viruses enter cells. Biologists even poke needles into cells without killing them. These examples show that molecular machines can reach and enter cells.
- Recognition. Antibodies and the tail fibers of the T4 phage - and indeed, all specific biochemical interactions - show that molecular systems can recognize other molecules by touch.
- Disassembly. Digestive enzymes (and other, fiercer chemicals) show that molecular systems can disassemble damaged molecules.
- Rebuilding. Replicating cells show that molecular systems can build or rebuild every molecule found in a cell.
- Reassembly. Nature also shows that separated molecules can be put back together again. The machinery of the T4 phage, for example, self-assembles from solution, apparently aided by a single enzyme. Replicating cells show that molecular systems can assemble every system found in a cell.

Thus, nature demonstrates all the basic operations that are needed to perform molecular-level repairs on cells, systems based on nanomachines will generally be more compact and capable than those found in nature. Natural systems show us only lower bounds to the possible, in cell repair as in everything else.

CELL REPAIR MACHINES

In short, with molecular technology and technical AI we will compile complete, molecular-level descriptions of healthy tissue, and we will build machines able to enter cells and to sense and modify their structures.

Cell repair machines will be comparable in size to bacteria and viruses, but their more-compact parts will allow them to be more complex. They will travel through tissue as white blood cells do, and enter cells as viruses do - or they could open and close cell membranes with a surgeon's care. Inside a cell, a repair machine will first size up the situation by examining the cell's contents and activity, and then take action. Early cell repair machines will be highly specialized, able to recognize and correct only a single type of molecular disorder, such as an enzyme deficiency or a form of DNA damage. Later machines (but not much later, with advanced technical AI systems doing the design work) will be programmed with more general abilities.

Complex repair machines will need nanocomputers to guide them. A micron-wide mechanical computer like that described earlier will fit in 1/1000 of the volume of a typical cell, yet will hold more information than does the cell's DNA. In a repair system, such computers will direct smaller, simpler computers, which will in turn direct machines to examine, take apart, and rebuild damaged molecular structures.

By working along molecule by molecule and structure by structure, repair machines will be able to repair whole cells. By working along cell by cell and tissue by tissue, they (aided by larger devices, where need be) will be able to repair whole organs. By working through a person organ by organ, they will restore health. Because molecular machines will be able to build molecules and cells from scratch, they will be able to repair even cells damaged to the point of complete inactivity.

Thus, cell repair machines will bring a fundamental breakthrough: they will free medicine from reliance on self-repair as the only path to healing.

To visualize an advanced cell repair machine, imagine it - and a cell - enlarged until atoms are the size of small marbles. On this scale, the repair machine's smallest tools have tips about the size of our fingertips; a medium-sized protein, like hemoglobin, is the size of a typewriter; and a ribosome is the size of a washing machine. A single repair device contains a simple computer the size of a small truck, along with many sensors of protein size, several manipulators of ribosome size, and provisions for memory and motive power. A total volume ten meters across, the size of a three-story house, holds all these parts and more. With parts the size of marbles packing this volume, the repair machine can do complex things.

But this repair device does not work alone. It, like its many siblings, is connected to a larger computer by means of mechanical data links the diameter of our arm. On this scale, a cubic-micron computer with a large memory fills a volume thirty stories high and as wide as a football field. The repair devices pass it information, and it passes back general instructions. Objects so large and complex are still small enough: on this scale, the cell itself is a kilometre across, holding one thousand times the volume of a cubic-micron computer, or a million times the volume of a single repair device. Cells are spacious.

Will such machines be able to do everything necessary to repair cells? Existing molecular machines demonstrate the ability to travel through tissue, enter cells, recognize molecular structures, and so forth, but other requirements are also important. Will repair machines work fast enough? If they do, will they waste so much power that the patient will roast?

The most extensive repairs cannot require *vastly* more work than building a cell from scratch. Yet molecular machinery working within a cellular volume routinely does just that, building a new cell in tens of minutes (in bacteria) to a few hours (in mammals). This indicates that repair machinery occupying a few percent of a cells volume will be able to complete even extensive repairs in a reasonable time - days or weeks at most. Cells can spare this much room. Even brain cells can still function when an inert waste called lipofuscin (apparently a product of molecular damage) fills over ten percent of their volume.

Powering repair devices will be easy: cells naturally contain chemicals that power nanomachinery. Nature also shows that repair machines can be cooled: the cells in our body rework themselves steadily, and young animals grow swiftly without cooking themselves. Handling heat from a similar level of activity by repair machines will be no sweat - or at least not too much sweat, if a week of sweating is the price of health.

All these comparisons of repair machines to existing biological mechanisms raise the question of whether repair machines will be able to *improve* on nature. DNA repair provides a clear-cut illustration.

Just as an illiterate "book-repair machine" could recognize and repair a torn page, so a cell's repair enzymes can recognize and repair breaks and cross-links in DNA. Correcting misspellings (or mutations), though, would require an ability to read. Nature lacks such repair machines, but they will be easy to build. Imagine three identical DNA molecules, each with the same sequence of nucleotides. Now imagine each strand mutated to change a few scattered nucleotides. Each strand still seems normal, taken by itself. Nonetheless, a repair machine could compare each strand to the others, one segment at a time, and could note when a nucleotide failed to match its mates. Changing the odd nucleotide to match the other two will then repair the damage.

This method will fail if two strands mutate in the same spot. Imagine that the DNA of three human cells has been heavily damaged - after thousands of mutations, each cell has had one in every million nucleotides changed. The chance of our three-strand correction procedure failing at any given spot is then about one in a million million. But compare five strands at once, and the odds become about one in a million million million, and so on. A device that compares many strands will make the chance of an uncorrectable error effectively nil. In practice, repair machines will compare DNA molecules from several cells, make corrected copies, and use these as standards for proofreading and repairing DNA throughout a tissue. By comparing several strands, repair machines will dramatically improve on nature's repair enzymes.

Other repairs will require different information about healthy cells and about how a particular damaged cell differs from the norm. Antibodies identify proteins by touch, and properly chosen antibodies can generally distinguish any two proteins by their differing shapes and surface properties. Repair machines will identify molecules in a similar way. With a suitable computer and data base, they will be able to identify proteins by reading their amino acid sequences.

Consider a complex and capable repair system. A volume of two cubic microns - about 2/1000 of the volume of a typical cell - will be enough to hold a central data base system able to:

- Swiftly identify any of the hundred thousand or so different human proteins by examining a short amino acid sequence.
- Identify all the other complex molecules normally found in cells.
- Record the type and position of every large molecule in the cell.

Each of the smaller repair devices (of perhaps thousands in a cell) will include a less capable computer. Each of these computers will be able to perform over a thousand computational steps in the time that a typical enzyme takes to change a single molecular bond, so the speed of computation possible seems more than adequate. Because each computer will be in communication with a larger computer and the central data base, the available memory seems adequate. Cell repair machines will have both the molecular tools they need and "brains" enough to decide how to use them.

Such sophistication will be overkill (overcure) for many health problems. Devices that merely recognize and destroy a specific kind of cell, for example, will be enough to cure a cancer. Placing a computer network in every cell may seem like slicing butter with a chain saw, but having a chain saw available does provide assurance that even hard butter can be sliced. It seems better to show too much than too little, if one aims to describe the limits of the possible in medicine.

SOME CURES

The simplest medical applications of nanomachines will involve not repair but selective destruction. Cancers provide one example; infectious diseases provide another. The goal is simple: one need only recognize and destroy the dangerous replicators, whether they are bacteria, cancer cells, viruses, or worms. Similarly, abnormal growths and deposits on arterial walls cause much heart disease; machines that recognize, break down, and dispose of them will clear arteries for more normal blood flow. Selective destruction will also cure diseases such as herpes in which a virus splices its genes into the DNA of a host cell. A repair device will enter the cell, read its DNA, and remove the addition that spells "herpes."

Repairing damaged, cross-linked molecules will also be fairly straightforward. Faced with a damaged, cross-linked protein, a cell repair machine will first identify it by examining short amino acid sequences, then look up its correct structure in a data base. The machine will then compare the protein to this blueprint, one amino acid at a time. Like a proofreader finding misspellings and strange characters, it will find any changed amino acids or improper cross-links. By correcting these flaws, it will leave a normal protein, ready to do the work of the cell.

Repair machines will also aid healing. After a heart attack, scar tissue replaces dead muscle. Repair machines will stimulate the heart to grow fresh muscle by resetting cellular control mechanisms. By removing scar tissue and guiding fresh growth, they will direct the healing of the heart.

This list could continue through problem after problem (Heavy metal poisoning? - Find and remove the metal atoms) but the conclusion is easy to summarize. Physical disorders stem from mis-arranged atoms; repair machines will be able to return them to working order, restoring the body to health. Rather than compiling an endless list of curable diseases (from arthritis, bursitis, cancer, and dengue to yellow fever and zinc chills and back again), it makes sense to look for the limits to what cell repair machines can do. Limits do exist.

Consider stroke, as one example of a problem that damages the brain. Prevention will be straightforward: Is a blood vessel in the brain weakening, bulging, and apt to burst? Then pull it back into shape and guide the growth of reinforcing fibers. Does abnormal clotting threaten to block circulation? Then dissolve the clots and normalize the blood and blood-vessel linings to prevent a recurrence. Moderate neural damage from stroke will also be repairable: if reduced circulation has impaired function but left cell structures intact, then restore circulation and repair the cells, using their structures as a guide in restoring the tissue to its previous state. This will not only restore each cell's function, but will preserve the memories and skills embodied in the neural patterns in that part of the brain.

Repair machines will be able to regenerate fresh brain tissue even where damage has obliterated these patterns. But the patient would lose old memories and skills to the extent that they resided in that part of the brain. If unique neural patterns are truly obliterated, then cell repair machines could no more restore them than art conservators could restore a tapestry from stirred ash. Loss of information through obliteration of structure imposes the most important, fundamental limit to the repair of tissue.

Other tasks are beyond cell repair machines for different reasons - maintaining mental health, for instance. Cell repair machines will be able to correct some problems, of course. Deranged thinking sometimes has biochemical causes, as if the brain were drugging or poisoning itself, and other problems stem from tissue damage. But many problems have little to do with the health of nerve cells and everything to do with the health of the mind.

A mind and the tissue of its brain are like a novel and the paper of its book. Spilled ink or flood damage may harm the book, making the novel difficult to read. Book repair machines could nonetheless restore physical - health" by removing the foreign ink or by drying and repairing the damaged paper fibers. Such treatments would do nothing for the book's *content,* however, which in a real sense is nonphysical. If the book were a cheap romance with a moldy plot and empty characters, repairs would be needed not on the ink and paper, but on the novel. This would call not for physical repairs, but for more work by the author, perhaps with advice.

Similarly, removing poisons from the brain and repairing its nerve fibers will thin some mental fogs, but not revise the content of the mind. This can be changed by the patient, with effort; we are all authors of our minds. But because minds change themselves by changing their brains, having a healthy brain will aid sound thinking more than quality paper aids sound writing.

Readers familiar with computers may prefer to think in terms of hardware and software. A machine could repair a computer's hardware while neither understanding nor changing its software

Such machines might stop the computer's activity but leave the patterns in memory intact and ready to work again. In computers with the right kind of memory (called "nonvolatile"), users do this by simply switching off the power. In the brain the job seems more complex, yet there could be medical advantages to inducing a similar state.

ANESTHESIA PLUS

Physicians already stop and restart consciousness by interfering with the chemical activity that underlies the mind. Throughout active life, molecular machines in the brain process molecules. Some disassemble sugars, combine them with oxygen, and capture the energy this releases. Some pump salt ions across cell membranes; others build small molecules and release them to signal other cells. Such processes make up the brain's metabolism, the sum total of its chemical activity. Together with its electrical effects, this metabolic activity underlies the changing patterns of thought. Surgeons cut people with knives. In the mid-1800s, they learned to use chemicals that interfere with brain metabolism, blocking conscious thought and preventing patients from objecting so vigorously to being cut. These chemicals are anesthetics. Their molecules freely enter and leave the brain, allowing anesthetists to interrupt and restart human consciousness.

People have long dreamed of discovering a drug that interferes with the metabolism of the entire body, a drug able to interrupt metabolism completely for hours, days, or years. The result would be a condition of biostasis (from *bio*, meaning life, and *stasis*, meaning a stoppage or a stable state). A method of producing reversible biostasis could help astronauts on long space voyages to save food and avoid boredom, or it could serve as a kind of one-way time travel. In medicine, biostasis would provide a deep anesthesia giving physicians more time to work. When emergencies occur far from medical help, a good biostasis procedure would provide a sort of universal first-aid treatment: it would stabilize a patient's condition and prevent molecular machines from running amok and damaging tissues.

But no one has found a drug able to stop the entire metabolism the way anesthetics stop consciousness - that is, in a way that can be reversed by simply washing the drug out of the patient's tissues. Nonetheless, reversible biostasis will be possible when repair machines become available.

To see how one approach would work, imagine that the blood stream carries simple molecular devices to tissues, where they enter the cells. There they block the molecular machinery of metabolism - in the brain and elsewhere - and tie structures together with stabilizing cross-links. Other molecular devices then move in, displacing water and packing themselves solidly around the molecules of the cell. These steps stop metabolism and preserve cell structures. Because cell repair machines will be used to reverse this process, it can cause moderate molecular damage and yet do no lasting harm. With metabolism stopped and cell structures held firmly in place, the patient will rest quietly, dreamless and unchanging, until repair machines restore active life. If a patient in this condition were turned over to a present-day physician ignorant of the capabilities of cell repair machines, the consequences would likely be grim. Seeing no signs of life, the physician would likely conclude that the patient was dead, and then would make this judgment a reality by "prescribing" an autopsy, followed by burial or burning.

But our imaginary patient lives in an era when biostasis is known to be only an interruption of life, not an end to it. When the patient's contract says "wake me!" (or the repairs are complete, or the flight to the stars is finished), the attending physician begins resuscitation. Repair machines enter the patient's tissues, removing the packing from around the patient's molecules and replacing it with water. They then remove the cross-links, repair any damaged molecules an structures, and restore normal concentrations of salts, blood sugar, ATP, and so forth. Finally, they unblock the metabolic machinery. The interrupted metabolic processes resume, the patient yawns, stretches, sits up, thanks the doctor, checks the date, and walks out the door.

FROM FUNCTION TO STRUCTURE

The reversibility of biostasis and irreversibility of severe stroke damage help to show how cell repair machines will change medicine. Today, physicians can only help tissues to heal themselves. Accordingly they must try to preserve the *function* of tissue. If tissues cannot function, they cannot heal. Worse, unless they are preserved, deterioration follows, ultimately obliterating structure. It is as if a mechanic's tools were able to work only on a running engine.

Cell repair machines change the central requirement from preserving *function* to preserving *structure*. Aswenoted in the discussion of stroke, repair machines will be able to restore brain function with memory and skills intact only if the distinctive structure of the neural fabric remains intact. Biostasis involves preserving neural structure while deliberately blocking function.

All this is a direct consequence of the molecular nature of

the repairs. Physicians using scalpels and drugs can no more repair cells than someone using only a pickax and a can of oil can repair a fine watch. In contrast, having repair machines and ordinary nutrients will be like having a watchmaker's tools and an unlimited supply of spare parts. Cell repair machines will change medicine at its foundations.

From Treating Disease To Establishing Health

Medical researchers now study diseases, often seeking ways to prevent or reverse them by blocking a key step in the disease process. The resulting knowledge has helped physicians greatly: they now prescribe insulin to compensate for diabetes, anti-hypertensives to prevent stroke, penicillin to cure infections, and so on down an impressive list. Molecular machines will aid the study of diseases, yet they will make understanding disease far less important. Repair machines will make it more important to understand health.

The body can be ill in more ways than it can be healthy. Healthy muscle tissue, for example, varies in relatively few ways: it can be stronger or weaker, faster or slower, have this antigen or that one, and so forth. Damaged muscle tissue can vary in all these ways, yet also suffer from any combination of strains, tears, viral infections, parasitic worms, bruises, punctures, poisons, sarcomas, wasting diseases, and congenital abnormalities. Similarly, though neurons are woven in as many patterns as there are human brains, individual synapses and dendrites come in a modest range of forms - if they are healthy.

Once biologists have described normal molecules, cells, and tissues, properly programmed repair machines will be able to cure even unknown diseases. Once researchers describe the range of structures that (for example) a healthy liver may have, repair machines exploring a malfunctioning liver need only look for differences and correct them. Machines ignorant of a new poison and its effects will still recognize it as foreign and remove it. Instead of fighting a million strange diseases, advanced repair machines will establish a state of health.

Developing and programming cell repair machines will

require great effort, knowledge, and skill. Repair machines with broad capabilities seem easier to build than to programme. Their programs must contain detailed knowledge of the hundreds of kinds of cells and the hundreds of thousands of kinds of molecules in the human body. They must be able to map damaged cellular structures and decide how to correct them. How long will such machines and programs take to be developed? Offhand, the state of biochemistry and its present rate of advance might suggest that the basic knowledge alone will take centuries to collect. But we must beware of the illusion that advances will arrive in isolation.

Repair machines will sweep in with a wave of other technologies. The assemblers that build them will first be used to build instruments for analyzing cell structures. Even a pessimist might agree that human biologists and engineers equipped with these tools could build and programme advanced cell repair machines in a hundred years of steady work. A cocksure, far-seeing pessimist might say a thousand years. A really committed nay-sayer might declare that the job would take people a million years. Very well: fast technical AI systems - a millionfold faster than scientists and engineers - will then develop advanced cell repair machines in a single calendar year.

A Disease Called "Aging"

Aging is natural, but so were smallpox and our efforts to prevent it. We have conquered smallpox, and it seems that we will conquer aging.

Longevity has increased during the last century, but chiefly because better sanitation and drugs have reduced bacterial illness. The basic human life span has increased little.

Still, researchers have made progress toward understanding and slowing the aging process. They have identified some of its causes, such as uncontrolled cross-linking. They have devised partial treatments, such as antioxidants and free-radical inhibitors. They have proposed and studied other mechanisms of aging, such as - clocks" in the cell and changes in the body's hormone balance. In laboratory experiments, special drugs and diets have extended the life span of mice by 25 to 45 percent.

Such work will continue; as the baby boom generation ages, expect a boom in aging research. One biotechnology company, Senetek of Denmark, specializes in aging research. In April 1985, Eastman Kodak and ICN Pharmaceuticals were reported to have joined in a $45 million venture to produce isoprinosine and other drugs with the potential to extend life span. The results of conventional anti-aging research may substantially lengthen human life spans - and improve the health of the old - during the next ten to twenty years. How greatly will drugs, surgery, exercise, and diet extend life spans? For now, estimates must remain guesswork. Only new scientific knowledge can rescue such predictions from the realm of speculation, because they rely on new science and not just new engineering.

With cell repair machines, however, the potential for life extension becomes clear. They will be able to repair cells so long as their distinctive structures remain intact, and will be able to replace cells that have been destroyed. Either way, they will restore health. Aging is fundamentally no different from any other physical disorder; it is no magical effect of calendar dates on a mysterious life-force. Brittle bones, wrinkled skin, low enzyme activities, slow wound healing, poor memory, and the rest all result from damaged molecular machinery, chemical imbalances, and mis-arranged structures. By restoring all the cells and tissues of the body to a youthful structure, repair machines will restore youthful health.

People who survive intact until the time of cell repair machines will have the opportunity to regain youthful health and to keep it almost as long as they please. Nothing can make a person (or anything else) last forever, of course, but barring severe accidents, those wishing to do so will live for a long, long time.

As a technology develops, there comes a time when its principles become clear, and with them many of its consequences. The principles of rocketry were clear in the 1930s, and with them the consequence of spaceflight. Filling in the details involved designing and testing tanks, engines, instruments, and so forth. By the early 1950s, many details were known. The ancient dream of flying to the Moon had became a goal one could plan for.

The principles of molecular machinery are already clear, and with them the consequence of cell repair machines. Filling in the details will involve designing molecular tools, assemblers, computers, and so forth, but many details of existing molecular machines are known today. The ancient dream of achieving health and long life has become a goal one can plan for.

Medical research is leading us, step by step, along a path toward molecular machinery. The global competition to make better materials, electronics, and biochemical tools is pushing us in the same direction. Cell repair machines will take years to develop, but they lie straight ahead.

They will bring many abilities, both for good and for ill. A moment's thought about military replicators with abilities like those of cell repair machines is enough to turn up nauseating possibilities. Later we will describe how we might avoid such horrors, but it first seems wise to consider the alleged benefits of cell repair machines. Is their *apparent* good *really* good? How might long life affect the world?

Chapter 2

Scientific Issues For Nanotechnology

VISION

Nanostructures are the entry into a new realm in physical and biological science. They are intermediate in size between molecular and microscopic (micron-size) structures. They contain a countable number of atoms, and are, as a result, uniquely suited for detailed atomic-level engineering. They are chameleons: viewed as molecules, they are so large that they provide access to realms of quantum behaviour that are not otherwise accessible; viewed as materials, they are so small that they exhibit characteristics that are not observed in larger (even 0.1 ìm) structures.

They combine small size, complex organizational patterns, potential for very high packing densities and strong lateral interactions, and high ratios of surface area to volume. Nanostructures are the natural home of engineered quantum effects. Microstructures have formed the basis for the technologies that support current microelectronics. Although microstructures are small on the scale of direct human experience, their physics is largely that of macroscopic systems. Nanostructures are fundamentally different: their characteristics—especially their electronic and magnetic characteristics—are often dominated by quantum behaviour.

They have the potential to be key components in information technology devices that have unprecedented functions. They can be fabricated in materials that are central to electronics, magnetics, and optics. Nanostructures are, in a sense, a unique state of matter—one with particular promise for new and potentially very useful products. Because they are small, nanostructures can be packed very closely together. Their high packing density has the potential to bring higher speed to information processing and higher areal and volumetric capacity to information storage. Such dense packing also is the cause of complex electronic and magnetic interactions between adjacent (and sometimes, nonadjacent) structures.

For many nanostructures, especially large organic molecules, the small energetic differences between their various possible configurations may be significantly shaped by those interactions. In some cases, the presence of surface interface material, with properties different from the nanostructures themselves, adds another level of complexity. These complexities are completely unexplored, and building technologies based on nanostructures will require in-depth understanding of the underlying fundamental science. These complexities also promise access to complex non-linear systems that may exhibit classes of behaviour fundamentally different from those of both molecular and microscale structures.

Exploring the science of nanostructures has become, in just a few years, a new theme common to many established disciplines. In electronics, nanostructures represent the limiting extension of Moore's law and classical devices to small devices, and they represent the step into quantum devices and fundamentally new processor architectures. In molecular biology, nanostructures are the fundamental machines that drive the cell— histones and proteosomes—and they are components of the mitochondrion, the chloroplast, the ribosome, and the replication and transcription complexes. In catalysis, nanostructures are the templates and pores of zeolites and other vitally important structures. In materials science, the nanometer length scale is the largest one over which a crystal can be made essentially perfect.

The ability to precisely control the arrangements of

impurities and defects with respect to each other, and the ability to integrate perfect inorganic and organic nanostructures, holds forth the promise of a completely new generation of advanced composites. Each of these disciplines has evolved its own separate view of nanoscience; the opportunities for integrating these views and for sharing tools and techniques developed separately by each field are today among the most attractive in all of science. Nanoscience is one of the unexplored frontiers of science. It offers one of the most exciting prospects for technological innovation. And if it lives up to its promise as a generator of technology, it will be at the centre of fierce international competition.

CURRENT SCIENTIFIC AND TECHNOLOGICAL ADVANCEMENTS

Nanoscience has exploded in the last decade, primarily as the result of the development of new tools that have made the characterization and manipulation of nanostructures practical, and also as a result of new methods for preparation of these structures.

Scaling Laws and Size-dependent Properties of Isolated Nanostructures

It is now well established that such fundamental properties as the melting temperature of a metal, the remanence of a magnet, and the band gap of a semiconductor depend strongly upon the size of the component crystals, provided they are in the nanometer regime. Almost any property in a solid is associated with a particular length scale, and below this length, the property will vary. For instance, the exciton diameter in a semiconductor may be tens or hundreds of nanometers, the distance between domain walls in a magnet may be hundreds of nanometers, etc.

This opens the prospect for creating a new generation of advanced materials with designed properties, not just by changing the chemical composition of the components, as has been done in the past, but by controlling the size and shape of the components. This creates great opportunities for fundamental science in condensed matter physics, solid state chemistry, materials science, electrical engineering, biology, and other disciplines.

Tools for Characterization

Scanning probe microscopies have revolutionized characterization of nanostructures, and development of new variants of scanning probe devices continues apace. Older tools, especially electron microscopy, continue to play essential roles. In biological nanoscience, the combination of X-ray crystallography and NMR spectroscopy offers atomic resolution structural information about structures as complex as entire virus particles.

Fabrication and Synthesis

Tremendous advances are currently occurring in the synthesis and fabrication of isolated nanostructures. These activities range from colloidal synthesis of nanocrystals to the growth of epitaxial quantum dots by strained layer growth. Related activities include the preparation of fullerenes, bucky tubes, and other one-dimensional nanostructures, as well as the growth of mesoporous inorganics. Increased activity in the nanoscale design of polymers is also occurring, including the development of dendrimers and complex block copolymers.

The techniques of molecular biology have made a very wide range of biological nanostructures readily available through cloning and over expression in bacterial production systems. While much has been accomplished in the growth of isolated nanostructures, work has only just begun in the use of self-assembly techniques to prepare complex and designed spatial arrangements of nanostructures. A parallel line of current activity in fabrication of patterned nanostructures rests on the extension of techniques highly developed in the field of microelectronics: photolithography, X-ray lithography, and e-beam lithography.

A number of recent developments in synthesis and fabrication offer the potential both to generate new types of structures, and, probably more importantly, to generate these structures at a fraction of the cost of techniques derived from microlithography.

Soft lithography, which uses molding, printing, and embossing to form patterned structures in plastics and glasses, has expanded the range of materials that can be used and has suggested routes to previously inaccessible three-dimensional structures.

Computation

Because nanostructures contain few atoms (at least relative to most materials), they are uniquely susceptible to high-level simulation using supercomputers. The capability to treat nanostructures with useful accuracy using computation and simulation will be invaluable both in fundamental science and in applied technologies.

Emerging Uses

Many clear applications for nanotools and nanostructures are already evident and are the targets of existing technology development programs:

- Giant magnetoresistance (GMR) materials have been introduced into commercial use with remarkable speed, and their acceptance suggests the importance of magnetic materials with nanometer-scale spin-flip mean free path of electrons.
- Numerous nanodevices and nanosystems for sequencing single molecules of DNA have been proposed; these structures, if successful, will be invaluable in the Human Genome Project and other large-scale genomics programs. Indeed, it seems quite likely that there will be numerous applications of inorganic nanostructures in biology and medicine, as markers.
- Similarly, there exists a range of ideas for high-density information storage, based, for example, on concepts such as nano-CDs and on nanostructured magnetic materials, including materials showing giant and tunneling magnetoresistive effects; these promise to provide future systems for memory with

ultrahigh densities.

- New types of components for information processors based on quantum mechanical principles (resonant tunneling transistors; single electron transistors; cellular automata based on quantum dots) are being explored actively at the level of research; these types of processors appear to fit well in the burgeoning field of quantum computation.
- New protective coatings, thin layers for optical filtering and thermal barriers, nanostructured polymers, and catalysts are already coming to the market. Nanostructured coatings are showing good corrosion/erosion resistance as possible replacements for the environmentally troublesome chromium-based coatings. Aerogels—highly porous, sponge-like materials with a three-dimensional filigree of nanostructures—have promise in catalysis and energy applications.

GOALS FOR THE NEXT 5-10 YEARS: BARRIERS AND SOLUTIONS

Numerous important areas require active research and development. The objectives of current research are to be able to understand the properties of isolated nanostructures; to make arbitrary structures with atomic-level precision; to do so rapidly, in large numbers, and inexpensively; and to design these structures to have desired properties using appropriate computer tools. Nanoscience is far from this objective, but it is moving rapidly in every component of the problem.

Fundamental Properties of Isolated Individual Nanostructures I

Individual nanostructures in isolation are the building blocks of nanotechnology. Individual structures are studied because we do not yet know the fundamental limits to the preparation of *identical* nanostructures and because each nanostructure can interact differently with its environment. Underlying the fundamental properties of nanostructures are two broad themes. First, the size-dependent properties of materials in the nanoscale regime are predicted to vary qualitatively according to scaling laws; comparison to these simple scaling laws remains an important activity. Second, the properties of isolated nanostructures have a significant statistical variation, fluctuating in time, and it is important to observe and understand these variations.

Many key questions relate to the structure, or arrangement of atoms, in a nanostructure. The relative stability of different structural phases is altered in the nanometer regime, affected by both kinetic and thermodynamic factors. Variations may arise for many reasons, including surface energies, absence of defects, or electronic quantum size effects. There is a compelling need to map out the kinetics and thermodynamics of phase transformations in nanostructures. For any picture of the physical properties of nanostructures to be complete, the structure of the surface must also be determined.

Due to their finite sizes, the structure and composition of the surfaces of nanostructures may have particular importance for their chemical and physical properties. The surfaces of nanostructures are likely to vary significantly from the well-known structures of bulk surfaces, and entirely new experimental techniques for measuring these reconstructions need to be developed. The ability not only to measure, but ultimately to systematically control, the surface and interior structures of nanoscale materials will be an ongoing field of research over the next decade. Recent developments have permitted the observation of optical, electrical, magnetic, chemical, thermal, mechanical and biological properties of isolated, individual nanostructures.

These techniques, which have facilitated developments such as single electron transistors, scanning probe microscopies, and single-molecule spectroscopy, have revolutionized our understanding of nanostructured materials. The early studies point the way to a long-term agenda: develop new probes of fundamental properties that can work with nanometer spatial resolution and ever-improved temporal resolution and sensitivity. Early work also reveals that nanoscale systems can be fluctuational by nature.

Much work is needed to understand their fluctuations and to learn what the fundamental limits are. As an example, an important question in quantum computation concerns whether it is possible to prepare well-defined superposition of quantum states in nanostructures without rapid dephasing.

Fundamental Properties of Ensembles of Isolated Nanostructures

Nanostructures may be used in a wide range of contexts; most of these are ones in which ensembles of nanostructures are assembled into a complex, functional arrangement. Many properties of nanoscale building blocks vitally depend upon the size, the shape, and indeed the precise arrangement of all the atoms within. Thus, a high priority must be placed upon understanding the fundamental limits to the preparation of identical nanostructures.

To this day, the processes of nucleation and growth are incompletely understood, and we do not know what is the largest number of atoms that can be assembled into a precisely defined molecular structure.

Even if a system is prepared according to an ideal chemical process, there inevitably will be variations. There is a need to understand further which desirable properties are retained or even simplified by averaging out such variations, and which properties are lost.

Assemblies of Nanoscale Building Blocks

The construction of functional assemblies of nanostructures depends upon a sound understanding of the intrinsic couplings between nanostructures. Charge separation and transport, tunneling, through-space electromagnetic coupling, and mechanical and chemical interactions between nanostructures need to be measured and theoretically described further than has been done to date. A combination of electron beam (and perhaps X-ray) lithography, scanning probe writing and fabrication, soft lithography, self-assembly, and catalytic growth together offer a rich menu of new and old fabrication techniques to use in patterning nanostructures.

With the exception of e-beam and X-ray lithography, these techniques are all early in their development cycle and have substantial promise for growth. The next decade will see these techniques developed and integrated into a suite of methods for the fabrication of nanostructures and nanosystems. The synthesis of bulk materials—colloids, magnetic structures, zeolites, bucky tubes and analogs, aerogels, and many others—is also an area where there is potential for great innovation. In most of these materials, the physics underlying their behaviour is understood only incompletely; there is an opportunity now to discover many new behaviours resulting from confinement of electrons or photons, from high structural perfection, from high ratios of surface to volume, or from some other aspect of small size.

In relevant but independent research in molecular biology, techniques of genetic manipulation, combined with technologies for protein production, provide the way to make small quantities of almost any protein of interest. The genome projects will extend this capability. Progress in biotechnology will increase the ease with which large quantities of genetic materials can be generated. A key issue is understanding the integration processes of various isolated nanostructures and assemblies of nanostructures.

Evaluation of Concepts for Devices and Systems

Among the most critical impediments to thinking seriously about nanostructure-based systems are the difficulties in understanding how they are to be interconnected and addressed and in understanding what kinds of new functions are achievable. For instance, in nanoelectronics, there are a number of solutions that have been suggested for some of these issues—ranging from using bucky tubes as nanowires to addressing individual components via ganged scanning probe devices, or even optically—but there has been almost no serious work directed toward the problems of systems fabrication.

Among the numerous questions that must be solved are:

what to use as wires; how to design and fabricate devices; and what the architectures of systems should be. Research is currently focused on exploring the fabrication and characterization of single devices at the level of single electron transistors or resonant tunneling devices.

There is much less effort (and certainly much less than is needed) devoted to asking fundamental questions about how electrical current is to be carried from a contact pad to a device in a densely packed array, fault- and defect-tolerant designs, device isolation, the operation of large arrays of cooperative devices, and other fundamental questions dealing with the problem of making an array of nanostructured quantum devices that performs some complex function. Similar to the nanoelectronics issues, resolving the aspects related to integration at nanoscale is essential in dispersions, nanocomposites, sensors, and other areas.

NANOMANUFACTURING

Developing techniques for fabricating nanostructures inexpensively in very large numbers—that is, manufacturing them—is an area that requires substantial effort: nanoscience will not be fully successful until it has provided the base for manufacturing technologies that are economically viable. It is probable that methods developed for microfabrication in the >100 nm size range will not work in the 20 nm range. It will thus probably be necessary to develop an entire new suite of manufacturing methods for nanostructures. There is reason to believe that self-assembly and soft lithography will be able to make substantial contributions to this important problem, but other fundamentally new methods will also be needed.

Connecting Nanoscience and Biology

One of the opportunities in basic science is to search for synergies between nanoscience in biology and nanoscience as developed in the contexts of computation and information science and solid-state physics and chemistry. There is no question that understanding the structure and function of biological nanostructures will stimulate fabrication of nonbiological materials; it is possible that biologically derived structures may also be useful in assembly of systems of nanodevices.

In return, nanofabrication can provide analytical tools for investigating biomolecules (in genomics, proteomics, and high through put screening for drug leads) as well as for exploring the interior structure and function of cells. One objective of nanoscience should be to build robust intellectual bridges between its currently scattered disciplinary components, but especially between nanoelectronics and molecular biology.

MOLECULAR ELECTRONICS

Molecular electronics offers an attractive opportunity for basic science in nanosystems. Organic molecules are probably the smallest systems that can be imagined for many possible functions, but they have the disadvantage, from the point of view of possible use in nanoelectronic systems, that they are usually poor conductors of electricity. A number of systems have been investigated in which experimental results suggest that organic molecules can act as molecular wires; there are hints that they may also serve as components in more complex systems. The science base in molecular electronics is very early in its development, and it is not yet clear whether organic molecules—even if they do conduct electricity—have the other properties they would need to be the core components of a large-scale nanoelectronics technology. Defining the real promise of molecular electronics should be an objective of research.

Nanostructures as Model Systems for Earth and Planetary Science

Nanoscale components are fundamental building blocks for solid state chemistry. Thus, further study of nanostructures can lead to improvements in our understanding of fundamental processes in earth and planetary science. Improved understanding of interstellar dust, the formation mechanisms of minerals, and the processes of weathering can all result from fundamental studies of nanostructures.

SCIENTIFIC AND TECHNOLOGICAL INFRASTRUCTURE

Tools for Synthesis and Fabrication, and for Characterization

To work in nanoscience, it is a prerequisite to be able to fabricate and characterize nanostructures: to make rabbit stew, one must first catch a rabbit. Making the important tools for fabrication and characterization available to user communities is an imperative; throughout the nation, research in nanoscience is still limited by limited access to tools. Certain instruments, especially electron microscopes, are sufficiently expensive that they should be operated within consortia; others, especially state-of-the-art scanning probe devices, should be distributed to qualifying individual research groups.

The character and cost of the facilities needed for fabrication generally depend on the technique being used. For the more expensive facilities—for example, high-resolution e-beam writers, good clean-room facilities, and mask-making facilities—a substantial, early investment is needed to prevent fabrication delays.

Flexible Research Structure

Nanoscience is an area in which there is no single way to do research: both singleinvestigator, peer-reviewed research and programmatic research involving groups of investigators from different disciplines and organizational structures have roles to play. It is important not to narrow the options for modalities of research support at this stage in the development of the field.

Training the Next Generation of Nanoscientists

Nanoscience has emerged into prominence only in the last 5-10 years. It is not supported in universities by existing departments: nanoscientists come from chemistry, physics, biology, electrical engineering, and others, and tend to think of themselves in terms of their historical disciplinary affiliations. Building educational programs that focus on nanoscience and nanotechnology as a distinct field—whether addressing device physics or structural biology—would provide a profound boost to the advancement of nanoscience by making it visible and intellectually attractive to the brightest young people and by promoting transfer of information between disciplines.

R&D INVESTMENT AND IMPLEMENTATION STRATEGIES

Because nanoscience is an area in which a range of research styles flourish, it is important to keep support distributed among a number of sources; it should not, at this juncture, be concentrated in one Federal agency. The National Science Foundation (NSF) and the Department of Defence Advanced Research Projects Agency (DARPA), for example, serve fundamentally different functions in nanoscience, and consolidation of these functions into one agency could slow progress in the field dramatically.

One of the challenges of an area of national importance but in which Government support is distributed to various agencies, is to coordinate Government-sponsored activities so that the whole is greater than the sum of the parts. NSF is the plausible agency to lead the coordination effort.

Reporting

To maintain a high focus on nanoscience/technology and to ensure careful reporting, NSF should be charged with providing an annual appraisal of the field to the Office of Management and Budget (OMB) and other interested parties such as various House and Senate committees, industrial users, etc.

Appropriate reporting and interagency discussion would help to maximize productive transfer of processes and materials among groups, regardless of the source of support.

Emphasis in Support

Nanoscience is still in its infancy. The emphasis in Federally funded research should be on precompetitive fundamental science and engineering. The object of the work should be to develop flexible, innovative methods of making, characterizing, applying, and manufacturing nanosystems. In nanoscience, as in many fields, technology transfer represents a difficult step.

In other fields of technology—especially biotechnology and information technology—the venture community has played a major role in technology transfer. Developing intellectual property and financial policies that make nanoscience an attractive investment would accelerate the development of commercial technologies.

PRIORITIES AND CONCLUSIONS

Interest in nanoscience is burgeoning, largely due to its enormous practical potential. Yet it is very clear that nanoscience has provided a new way of thinking that has the potential to promote truly exciting developments in fundamental science as well. Long-standing questions in condensed matter physics and chemistry, in biological science, in materials science, and in mechanical engineering will receive renewed attention as a result of the exciting developments in nanotechnology.

Priorities

- Build a broad programme of R&D in nanoscience that will include research universities, relevant industry, and some of the national laboratories. Important public policy objectives in this field should include building a "community" focused on nanoscience/ technology and providing stable support for this community at a level high enough to allow the participating groups to reach a critical mass.
- Develop other mechanisms for bridging the gaps between communities interested in nanoscience, including both the physical and biological sciences.
- Develop policies explicitly designed to attract large companies as participants in programs of Federally funded groups. Without the large-company participation, technology development programs in

nanoelectronics will probably fail at the stage of research planning and product definition.

- Develop a strategy for informal coordination of R&D among participating Federal agencies, centered in NSF.
- Maintain an active series of reports to OMB and Congress, both to aid in educating policymakers about the progress, opportunities, and failures of the field, and to provide a general education about nanoscience at senior levels in the Government.
- Provide Federal funds to push science that seems to offer the potential of developing into profitable technology rapidly to the point of manufacturable prototypes using focused, DARPA-style programs.
- Address the problems of public perception of threats from nanoscience with active programs to reduce any possible threats and educate the public.

EXAMPLES OF CURRENT ACHIEVEMENTS AND PARADIGM SHIFTS

Quantum Dot Formed by Self-assembly (Ge "pyramid")

Figure is a scanning tunneling microscope (STM) image of a pyramid of germanium atoms on top of a silicon surface.

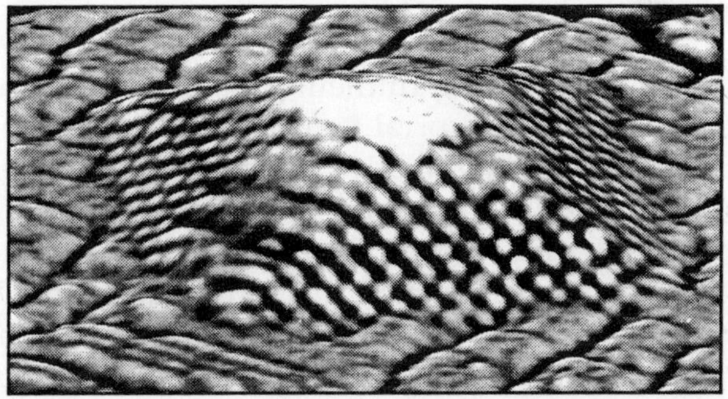

Fig. STM Image of Quantum dot Formed by Self-assembling

The pyramid is ten nanometers across at the base, and it is actually only 1.5 nanometers tall (the height axis in the image has been stretched to make it easier to see the detail in the faces of the pyramid). Each round-looking object in the image is actually an individual germanium atom.

The pyramid forms itself in just a few seconds in a process called "self-assembly." If the proper number of germanium atoms is deposited onto the correct type of silicon surface, the interactions of the atoms with each other causes the pyramids to form spontaneously. The propensity of some materials to self-assemble into nanostructures is currently a major area of research.

The intent is to learn how to guide or modify self-assembly to get materials to form more complex structures, such as electronic circuits. Manufacturing processes based on guided self-assembly of atoms, molecules, and supramolecules promise to be very inexpensive. Instead of requiring a multibillion-dollar manufacturing facility, electronic circuits of the future may be fabricated in a beaker using appropriate chemicals, and yet they may be many thousands to millions of times more capable than current chips.

Just twenty years ago, few scientists even dreamed it would be possible to see such a detailed picture of the atomic world, but with the advent of new measuring tools, the scanning probe microscopes developed in the mid-1980s, seeing atoms is now an everyday occurrence in laboratories all over the world.

QUANTUM CORRAL

Figure is a scanning tunneling microscope image of a "quantum corral." The corral is formed from 48 iron atoms, each of which was individually placed to form a circle with a 7.3 nanometer radius. The atoms were positioned with the tip of the tunneling microscope. The underlying material is pure copper. On this copper surface there are a group of electrons that are free to move about, forming a so-called "two-dimensional electron gas." When these electrons encounter an iron atom, they are partially reflected.

The purpose of the corral is to try to trap, or "corral" some of the electrons into the circular structure, forcing the trapped electrons into "quantum" states. The circular undulations in the interior of the corral are a direct visualization of the spatial distribution of certain quantum states of the corral. Experiments such as this give scientists the ability to study the physics of nanometer-scale structures and to explore the potential application of these small structures to any of a number of purposes.

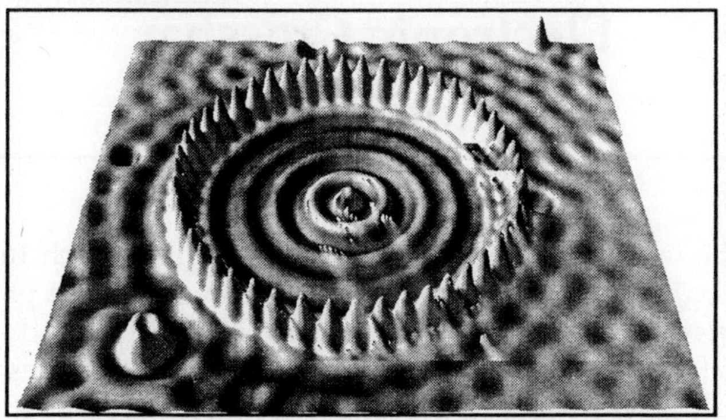

Fig. STM Image of a "quantum corral"

MAGNETIC BEHAVIOUR AT NANOSCALE

Figure is a transmission electron microscope (TEM) image of an elegant example of natural nanotechnology that occurs in magnetotactic bacteria. These are bacteria that contain within them a "compass" that allows them to move in a particular magnetic direction. The compass consists of a series of magnetic nanoparticles arranged in a line.

Chapter 3

Electronic Crossbar Memory Circuits

The fabrication procedures for molecular switch tunnel junction (MSTJ) device at a single device level and also proved that bistable rotaxane plays a crucial role in a decent conductance switching. The rotaxane molecular switches hold several advantages over more traditional switching components such as:

- Ferroelectric
- Ferromagnetic materials.

However, a key application of these molecular switches is related to the extreme scaling of electronic circuitry to near molecular dimensions: since conductance switching within an MSTJ originates from the electrochemically driven molecular mechanical isomerization of the molecules, the switching relies purely on individual molecular properties. In both molecular dynamics (MD) and electrochemical investigations, the two molecular co-conformers are characterized by different HOMO-LUMO gaps, and therefore different tunneling probabilities.

By contrast, the switching of solid state materials such as ferroelectrics involves altering the polarization state of the crystallographic lattice upon the application of an external electric field. This polarization disappears for domains below a certain critical size. An analogous phenomennon is the transition from ferromagnetic behaviour to superparamagnetic behaviour for magnetics, as the size of the ferromagnetic material is reduced. Therefore, these devices have a critical limitation in the scale-down. Second, solid state material switching depends upon a field-driven nucleation process and can be statistical in nature, especially for small crystallographic domain sizes.

The molecular switches discussed in this thesis, by contrast, switch based upon electrochemical processes. These are current and voltage driven, and depend upon molecular properties such as redox potentials, molecular orbital energies, etc. Therefore, switching voltages for the ferroelectric device could vary from junction to junction, or during many cycles.

In particular, at a smaller dimension, the reliability issues are expected be more serious. When the solid state materials are integrated into a 2D crossbar circuit that is a main architecture of the memory devices described here, their irregular switching characteristic will cause an additional problem, that is, 'a half select issue'.

In the crossbar memory circuit, a switching voltage, VA, is split into two components, +½ VA and -½VA, which are then applied to the top and bottom electrodes that define a designated cross-junction.

As all of the junctions are interconnected in a crossbar circuit, every junction is subject to at least some field. For the case of the solid state materials, the field generated by ± ½ VA is occasionally sufficient to perturb the state of the nucleation event. This half-select problem is considered as a generic problem for the field-poled devices that function within a 2D crossbar circuit. Taking advantage of the switching molecules, the next challenge is to integrate those switching molecules and electrodes into fully functional circuits, patterned at nanometer dimensions.

When such a memory circuit is combined with other nanoscale functional circuits such as logic and routing, computing at nanometer dimensions, which is currently unachievable with standard CMOS technology, could be realized. Entire circuits for nanoscale computing were proposed as conceptualized in figure. While efforts in my research group have focused on developing and integrating the various components of this nano-computer, this chapter will address fabrication and testing of crossbar memory circuits exclusively.

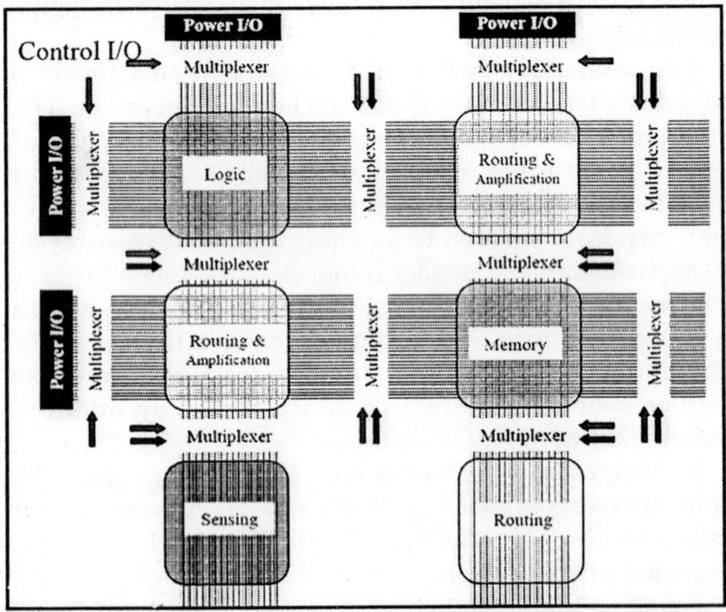

Fig. A nanoscale molecular computational platform. Tranditional computing functions are coupled to non-traditional elements, including sensors, actuator, etc., and are illustrated as individual tiles in a mosaic-like architecture. Muliplexers and demultiplexers control communication between various functions and provide the user interfaces.

As presented in figure, the proposed computing platform is based on the crossbar architecture. The crossbar geometry provides a promising architecture for nanoelectronic circuitry. The crossbar is tolerant of manufacturing defects – a trait that becomes increasingly important as devices approach macromolecular dimensions and non-traditional (and imperfect) fabrication methods are employed.

For example, Teramac had nearly a quarter million hardware defects and yet could be configured into a robust computing machine. The crossbar is a periodic array of crossed wires, similar to a two-dimensional crystal, implying that non-traditional methods can be employed for its construction. Finally, the crossbar is the highest density, two-dimensional digital circuit for which every device can be independently addressed. This attribute enables the circuit to be fully tested for manufacturing defects and to be subsequently configured into a working circuit. For these reasons, my research group has tried to utilize the crossbar architecture for memory circuits. The progress in this crossbar memory project has been made both in the switching molecules and in the contacting electrodes: more robust rotaxanes with higher on/off ratios have been rationally designed and synthesized. At the same time, as an attempt to constitute a circuit with higher density, various electrode-patterning techniques have been developed and tested.

In practice, the total number of bits fabricated within a single crossbar circuit increased from 16 to 64 to 4,500 to 160,000 bits as the main electrode pattering techniques were improved from photo-lithography to electron-beam lithography (EBL) to recently developed nanowire array technique, and as methods to increase the compatibility of our nanofabrication procedures with the molecular switch components were improved. Notably, the resolution of the nanowire array that has been developed by my research group is far beyond that of the conventional EBL.

Hence, the memory circuits utilizing this nanowire array technique set a remarkable landmark in memory bit density. In this chapter, we will first describe the nanowire array technique, so called superlattice nanowire pattern transfer (SNAP) method, and then focus on fabrication and testing of 160 kbit crossbar memory circuits based on the SNAP technique and rotaxanes. The fabrication of the 160 kbit memory circuits at a bit density of 10^{11} bits/cm^2 was totally nontrivial. At this point, reviewing the history of molecular memory projects in my research group provides guidance for understanding our efforts toward device miniaturization, as well as understanding difficulties of the fabrication procedure.

Until reaching this unexplored bit density, many scientists from several groups have contributed to different components of the circuits at each stage: In conjunction with Hewlett-Packard group, my research group initially proposed the concept of the defect-tolerant crossbar architecture. Since then, significant progresses in achieving the actual molecule-based devices were initiated by the former postdocs in my research group, Dr. Collier, Dr. Wong and Dr. Luo. They optimized conditions for many of the key fabrication steps, including monolayer deposition by the Langmuir-Blodgett technique and metal deposition to form the top electrical contact to the molecules.

They also established rational electrical measurement schemes such as the remnant molecular signature scan and the temperature-dependent volatility scan to test the switching and the activated nature of the molecular electronic switching mechanism.

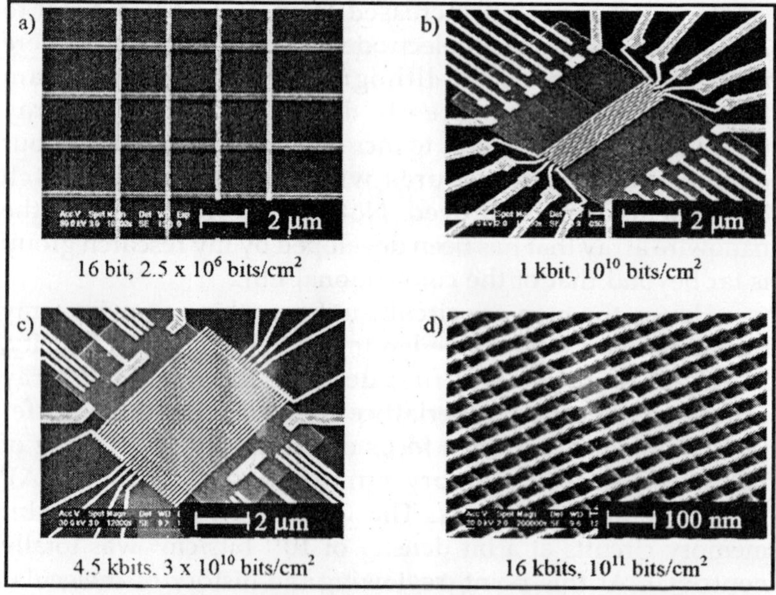

Fig. A series of crossbar molecular electronic memory circuits. These circuits are arranged a-d in accordance with the chronology of their fabrication. The total number of bits and the memory density of each memory circuit were denoted below scanning electron micrograph (SEM) pictures.

They also demonstrated relative robust molecular switches that could be cycled > 1,000 times. Dr. Luo developed next-generation fabrication procedures for the memory circuits based on the EBL-defined electrodes.

He demonstrated that the molecular switching in EBL-defined circuits is still very robust, and he developed a number of procedures for the integration of Si SNAP nanowire bottom electrode arrays with EBL defined top electrodes. For my part, was teamed up with another group member, Jonathan Green, to achieve the large-scale (160,000 bit) SNAP nanowire-based memory circuits at extreme density (10^{11} bitscm^{-2}).

SUPERLATTICE NANOWIRE PATTERN TRANSFER (SNAP) METHOD

As described in the previous subchapter, the scaling advantage of a rotaxane molecular electronic switch would be best illustrated only when electrodes with molecular dimensions are integrated together to form the junction sandwiching the molecule.

For this and other reasons, my research group has developed unique nanowire array fabrication technique called superlattice nanowire pattern transfer (SNAP).

For the SNAP method, the layer structure of a GaAl/AlxGa(1- x)As superlattice in which each layer is grown under atomic-level control, is translated into a variety of metals or silicon.

The width and pitch of final nanowire array made of metals and/or silicon are defined by the initial superlattice film widths and spacings. With the SNAP method, my research group has demonstrated the fabrication of silicon nanowire arrays in which each wire is about 8 nm wide and at a pitch of about 16 nm. Also, nanowire arrays containing up to 1,400 nanowires have been demonstrated. Moreover, in comparison to other nanowire growth methods, such as the vapour-liquid-solid growth method most fully developed in the Lieber group, the SNAP method has no limitation in length of the nanowires within arrays. A few millimeter long nanowire arrays are routinely produced.

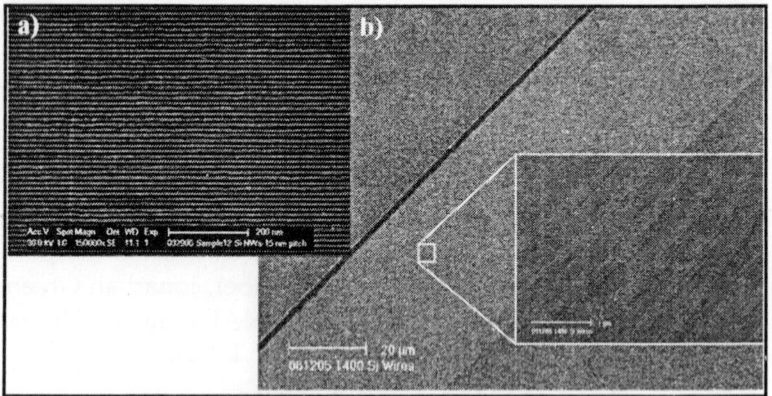

Fig. SEM images of Si nanowire arrays of (a) 15 nm pitch and (b) 1400 wires.

Here,we briefly go through the SNAP procedures. More detailed procedures of the SNAP method are described in the thesis of Jonathan Green, who was also involved in the project as a leading member.

First, a wafer containing the superlattice was diced into small pieces, which are referred to as masters. When turned on its side, a master is largely a GaAs wafer, with the top edge of the wafer containing the superlattice. When viewed from the edge, this superlattice structure is like a club sandwich, with alternating layers of GaAs and AlxGa(1-x)

As films substituting for the meat and bread layers in the sandwich. It is this superlattice edge that provides the initial template for the nanowires. This edge is first cleaned carefully in a class 1000 clean room using methanol and gentle swabbing, so that it is clean by eye when viewed under an optical microscope. Once the cross-section turns out to be completely dust-free, a dilute mixture of buffered hydrofluoric acid (15 ml of 6:1 buffered oxide etchant, 50 ml of H_2O) is used to selectively etch the AlxGa(1-x)As, thus forming a comb-like structure as shown in figure. Conversely, the GaAs could be selectively etched to form a complementary comb-like structure. Next, about 10 nm of Pt layer was deposited onto the superlattice side at about 45° tilted angle using electron beam evaporation of a Pt target.

This angle could be varied to yield some control over the nanowire width. In preparing silicon substrates (using silicon-oninsulator wafers), the substrates were cleaned by rinsing with a series of solvents until the surface becomes completely dust-free. A mixture of epoxy and poly (methyl methacrylate) (PMMA) (10 drops of epoxy and 1 drop of curing agent, 0.18 g of 6 % (in weight) PMMA in chlorobenzene, additional 10 ml of chlorobenzene) was spincast onto prepared substrates at 8000 rpm. Pt deposited masters were dropped onto the epoxy coated substrates so that the superlattice side contacted and adhered to the epoxy layer.

The epoxy was cured around 130 °C for half an hour or so. The substrates were then dipped into an etching solution for removing GaAs masters, but leaving behind the Pt nanowires, which were epoxy-adhered to the substrate. After ~ 4 hr of wet-etching, the GaAs masters were peeled off and the Pt nanowire array structure remained on the substrate. Once the quality of the Pt nanowire array was confirmed by SEM, the Pt pattern was transferred into the underlying silicon-on insulator substrate by reactive ion etching (RIE) (CF_4:He = 20:30 sccm, 5 mTorr, 40 W, ~ 4 min for 33 nm Si layer). Upon removing Pt by aqua resia (HCl:HNO_3 = 2:1 in volume), the Si nanowire array structure was complete.

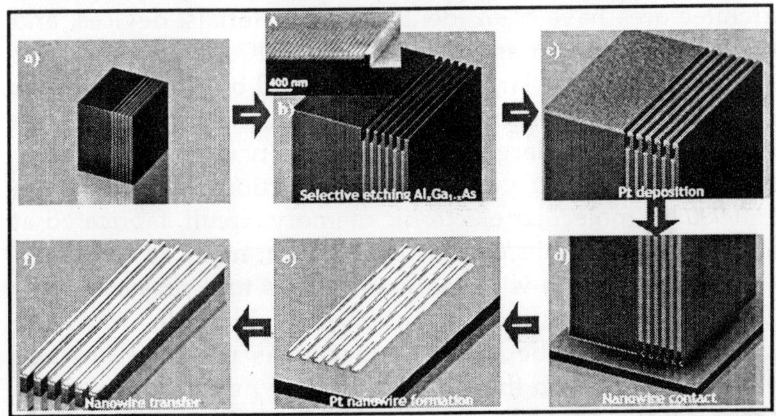

Fig. SNAP Process Flow.

(a) The wafer containing the superlattice was diced into

small pieces and the superlattice side was cleaned thoroughly. (b) AlxGa(1-x)As was selectively wet-eched. (c) A Pt layer was deposited onto the superlattice side by electron beam evaporation. (d) Masters were dropped onto the epoxy coated substrates. (e) The superlattice was removed by a wet-etch, leaving the Pt nanowirearray structure on the substrates. (f) The Pt nanowire array structure was transferred to form an array of aligned and high aspect ratio silicon nanowires via RIE.

160 KBIT MOLECULAR ELECTRONIC MEMORY CIRCUITS: OVERVIEW

The primary metric for gauging progress in the various semiconductor integrated circuit (IC) technologies is the spacing, or pitch, between the most closely spaced wires within a dynamic random access memory (DRAM) circuit. Modern DRAM circuits have 140 nanometer (nm) pitch wires and a memory cell size of 0.0408 square micrometers (ìm^2). Improving IC technology will require that these dimensions decrease over time. However, by year 2013 a large fraction of the patterning and materials requirements for constructing IC technologies are currently classified as having 'no known solution'. Nanowires, molecular electronics, and defect tolerant architectures have been identified as materials, devices, and concepts that might assist in continuing IC advances.

This belief has largely been bolstered by single device or small circuit demonstrations. The science of extending such demonstrations to large scale, high density circuitry is largely undeveloped. In this and following sections, we describe a 160,000 bit molecular electronic memory circuit, fabricated at a density of 10^{11} bits/cm^2 (pitch = 33 nm; memory cell size = 0.0011 ìm^2), which is roughly analogous to a projected year 2020 DRAM circuit. A monolayer of bistable, rotaxane molecules described earlier served as the data storage elements. Although the circuit had large numbers of defects, those defects could be readily identified through electronic testing and isolated using software coding.

The working bits were then configured to form a fully

functional random access memory circuit for storing and retrieving information. A few groups have reported on non-lithographic methods for fabricating crossbar circuits, but most methods are not yet feasible for fabricating more than a handful of devices. Furthermore, the assembly of nanowires into narrow pitch crossbars without electrically shorting adjacent nanowires remains a challenge. Despite these challenges, my research group developed the SNAP method for producing ultra-dense, highly aligned arrays of high-aspect ratio metal or semiconductor NWs containing up to 1400 NWs at a pitch as small as 15 nm.

The procedures for this SNAP method were described in the previous subchapter. For constituting ultra-dense memory circuits whose density is far beyond what is possible with current CMOS technology, combined these patterning methods and extremely scalable rotaxane switches, along with the defect-tolerance concepts learned from Teramac. we constructed and tested a memory circuit at extreme dimensions: the entire 160,000 bit crossbar is approximately the size of a white blood cell (~13×13 ìm^2). At each cross-point of nanowire array, only several hundreds of rotaxanes were incorporated.

160 KBIT MOLECULAR ELECTRONIC MEMORY CIRCUITS: FABRICATION FLOW

A bottom-up approach was the key to the successful fabrication of this memory. This approach both minimized the number of processing steps following deposition of the molecular monolayer, as well as protected the molecules from remaining processing steps. In the following paragraphs, we describe the nanofabrication procedures utilized to construct the memory circuit. Our 160,000 junction crossbar memory consists of 400 Si nanowire (NW) bottom electrodes of 16 nm width and 16.5 nm half-pitch, crossed with 400 Ti NW top electrodes of the same dimensions, and with a monolayer of bistable rotaxane molecules sandwiched in between.

My research group has previously reported on using the SNAP technique to fabricate highly ordered arrays of 150 metal and Si NWs. For this work, the SNAP technique was extended to create 400 element NW arrays of both the bottom and top electrode materials, and so was the primary patterning method for achieving the 1011 cm^{-2} bit density of the crossbar. An overview of the process flow used to fabricate the memory is shown in figure.

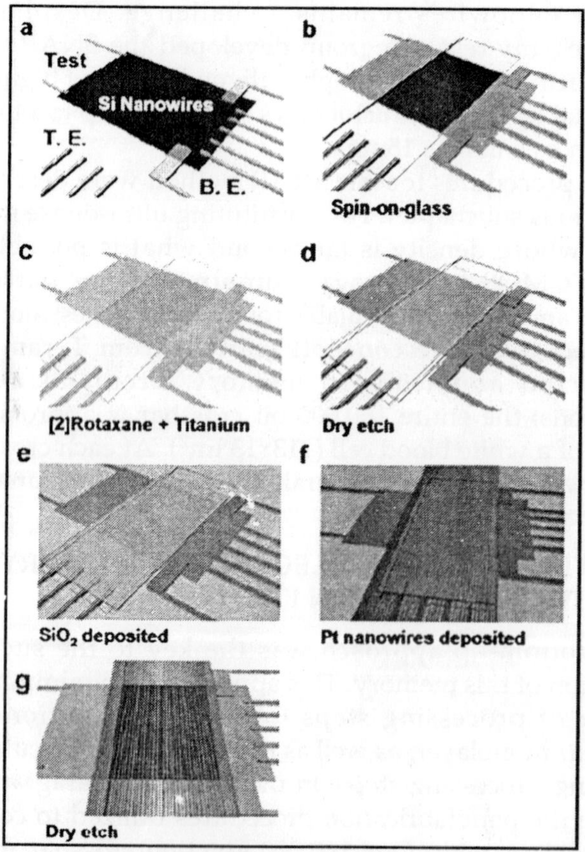

Fig. The Process Flow for Preparing the 160 kbit Molecular Electronic Memory Circuit at 10^{11} bits/cm^2.

(a) SNAP-patterned SiNW bottom electrodes are electrically contacted to metal electrodes. (b) The entire circuit is coated with SiO_2 (using spinon- glass (SOG)) and the active memory region is exposed using lithographic patterning followed by dry etching. (c) The bistable rotaxane Langmuir monolayer is deposited on top of the Si NWs and then protected via the deposition of a Ti layer. (d) The molecule/Ti layer is etched everywhere except for the active memory region. (e) A SiO_2 insulating layer is deposited on top of the Ti film. (f) An array of top Pt NWs is deposited at right angle to the bottom Si NWs using the SNAP method. (g) The Pt NW pattern is transferred, using dry etching, to the Ti layer to form an array of top Ti NW electrodes, and the crossbar structure is complete.

Preparation of and Contact to the Bottom Si Nanowire Electrodes

The Si NW array was fabricated as described previously. The starting wafer for the Si NWs was a 33 nm thick phosphorous doped ($n=5x10^{19}$ cm^{-3}) silicon-on-insulator (SOI) substrate with a 250 nm thick buried oxide (Simgui, Shanghai, China). An array of Pt NWs was transferred onto this substrate using the SNAP method, and reactive ion etching was used to transfer the Pt NW pattern to form a ~2 millimeter long array of Si NWs.

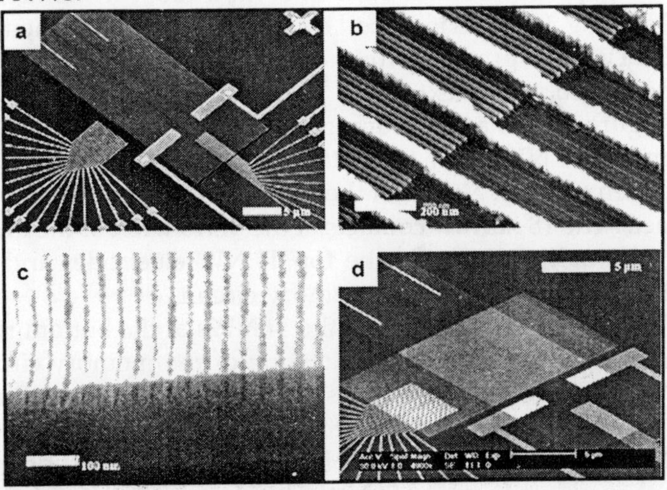

Fig. Scanning Electron Micrographs of the Nanowire Crossbar Memory Fabrication Process.

The Pt NWs were then removed, and the Si NW array was sectioned into a 30 ìm long region. Electrical contacts to these bottom Si NWs, as well as contacts that are intended for the top Ti NWs were defined at this point using standard electron-beam lithography (EBL) patterning and electron-beam evaporation to produce electrodes consisting of a 15 nm Ti adhesion layer followed by a 50 nm thick Pt electrode.

Immediately prior to metal evaporation, the Si NWs were cleaned using a gentle O_2 plasma (20 standard cubic centimeters per minute (sccm), 20 milliTorr, 10 Watts, 30 seconds) followed by a 5 second dip in an NH_4F/HF solution. After lift-off, the chip was annealed at 450 °C in N_2 for 5 min to promote the formation of ohmic contacts.

(a) A 30 micrometer long section of the SiNWs and its electrical contacts to metal leads were defined by electron-beam lithography (EBL). (b) Each electrode defined by EBL is about 70 nm wide contacting 2 ~ 4 NWs. This image illustrates that the intrinsic patterning of nanowire crossbar is beyond lithographic limits. (c) Progress-check of SOG window etching over the active memory region. This image verifies that SOG fills the gap of NWs and SEM is a valid tool for monitoring SOG etching. (d) SOG is etched by RIE over the active memory region. Detailed processes for monitoring this etching progress are described in the text.

Figure shows an SEM image of the device at the stage in which the Si NWs and all of the external electrical contacts have been created. Note that there are four sets of EBL defined contacts. The 18 narrow contacts at the bottom left of the image will eventually connect to the top Ti NW electrodes and are used for testing of the final memory circuit.

The 10 narrow contacts to the Si NWs at the bottom right of the image are also used for testing of the memory circuit. Finally there are two narrow test electrodes at the top left and two wide electrodes at the bottom right. The wide electrodes contact about 2/3 of all the Si NWs and serve dual functions. First, they ground unused Si NWs during memory testing (this procedure approximates how a fully multiplexed crossbar circuit would be utilized). Second, when used in conjunction with the two narrow test-electrodes on the opposite side of Si NW array, they enable testing of the conductivity of the Si NWs throughout the fabrication processes.

This testing procedure provided invaluable feedback for finely tuning and tracking many of the fabrication processes. Once these various contacts were established, robust Si NW conductivity was confirmed via current vs. voltage measurements. If the Si NWs were measured to be poor conductors (a very infrequent occurrence), the chip was discarded. The device was then planarized using an optimized spin-on-glass (SOG) procedure.

This planarization process was critical because the SOG not only protects Si NWs outside of the active memory region from damage that can arise during subsequent processing steps, but it also prevents evaporated Ti from entering the gaps between the Si NWs where it would be extremely difficult to remove. Due to the extremely narrow gap between the Si NWs, this SOG step was performed in a vacuum condition: For the first generation of the devices, the SOG was spincoated at atmospheric pressure. However, the atmospheric spin-coating did not allow the gaps between the Si NWs to be filled completely with the SOG. The SOG penetrated only to the upper spacing of the trenches.

For the complete filling, the process was done in a vacuum condition. The substrate containing the Si NWs was placed in a small glass container covered by a rubber stopper. A needle connected to a syringe and to a diffusion pump was plugged through the rubber stopper to employ a vacuum condition. During this vacuum process, the SOG was transferred to the container via another syringe. As soon as the SOG was sucked into the container by the vacuum and therefore the substrate was covered by the SOG, the substrate was taken out immediately for a spin-coating (~ 5000 rpm, 30 sec). Before starting the planarization steps described so far, all the glasswares including the container were cleaned very carefully because even a small dust particle could ruin the device.

Especially, the top SNAP NWs process requires very clean and flat surfaces in a several millimeters range. In some cases, some dust particles appeared during the vacuum transfer despite the careful preparation.

For that case, the SOG was stripped by methanol and then the substrates were cleaned intensively by spraying methanol onto the substrates followed by blowing the dust particles off with nitrogen gas repeatedly.

Upon confirming that the surface is completely dust-free, SOG was spincoated again at an atmospheric condition. For this second SOG spin-coating, the vacuum condition was not necessary: As mentioned above, the trenches between Si NWs are not usually completely filled with the SOG if the substrate is spin-coated directly at atmospheric pressure.

However, the second spin-coating performed at atmospheric pressure fills the trenches completely with the SOG, as indicated by figure.

Next, SOG layer thinned down globally using a CF4 plasma (20 standard cubic centimeters per minute (sccm), 10 milliTorr, 40 Watts). This etching was monitored periodically by ellipsometer and continued until the SOG layer became about 50 nm thick according to the ellipsometer. This final thickness is very critical because it affects the ensuing top SNAP and Ti layer dry-etching steps significantly. The detailed reasons are described in the paragraphs dedicated to those steps. After globally thinning the SOG layer, an opening in photo resist was lithographically defined over the Si NWs and the tips of the 18 EBL defined contacts.

The SOG was then further etched until the tops of the underlying Si NWs were exposed. This step was monitored by periodically measuring the Si NW conductivity using the test electrodes.

The majority of the dopant atoms in the Si NWs lie within the top 10 nm of the NWs. This feature means it is very straightforward to etch back the SOG without thinning the Si NWs, since the conductivity of the NWs is very sensitive to their thickness.

At this stage the entire memory circuit is under SOG (and thus electrically isolated from any further top processing) except for the lithographically defined opening over the Si

NWs and the 18 contacts. This opening defines the active memory region.

DEPOSITION OF MOLECULES AND TOP ELECTRODE MATERIALS

A monolayer of bistable rotaxane switches was prepared by Langmuir-Blodgett techniques and transferred onto the device as reported previously. For the rotaxane used herein, the Langmuir-monolayers were prepared on an aqueous (18 MÙ H_2O) subphase of Langmuir-Blodgett (LB) trough (Type 611D, Nima Technology, Coventry, UK). Before the trough was filled with the subphase, all the parts in the trough including compression barriers were cleaned very carefully by wiping with chloroform soaked wipes. Once the parts in the trough were wiped thoroughly, the filtered water was poured until the water level reached the compression barrier. From this point, the quality of the subphase was monitored by a brewster angle microscope (BAM).

For further cleaning, the subphase was compressed to an area of about 50 cm^2 and then the surface of the subphase was sucked by a glass pipette connected to a pump to remove dust particles floating on the subphase surface. As the compression and cleaning processes were repeated, the number of dust particles decreased and eventually, no dust particle was observed in the BAM image. Then, the barrier was moved back to the open position (~ 245 cm^2) and the prepared rotaxane solution was dropped onto the subphase via a syringe.

The rotaxanes were prepared in a chloroform solution right before the transfer. After about 30 minutes of the chloroform evaporation, the barrier compression began at a rate of 5 cm^2/min. Once the surface pressure reached the target pressure (ð = 30 mN/m), the surface pressure was fixed and the substrate started to be pulled out at a rate of 1 mm/min. When the entire substrate was pulled out of the subphase, the step for the preparation of the Langmuir-monolayer was complete. 20 nm of Ti was then evaporated over the entire device.

This Ti layer serves to protect the molecules from further

top processing. Using photolithographic techniques and BCl_3 plasma etching (10 sccm, 5 mTorr, 30 Watts), the molecule/Ti layer was then everywhere removed except for the memory active region where electrical contact to the underlying Si NWs is made. Next, a thin SiO_2 layer (~ 15 nm) was deposited over the entire substrate to isolate the EBL defined electrodes from the Pt NWs deposited in the next step. Remember that the SOG layer was about 50 nm thick after the SOG global etching step as described in the previous paragraph so that the EBL defined electrodes as thick as 65 nm were exposed until the SiO_2 layer deposition.

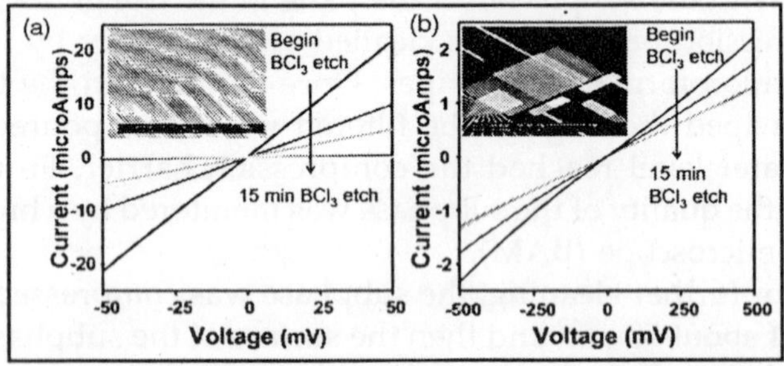

Fig. Conductance Monitoring During the Ti Layer Etching.

(a) Crossconductance measurements between electrical contacts to the top nanowire array were performed to monitor the Ti layer etching. When the current drops to sub-10 nanoAmps, the top Ti electrodes are separated. The inset SEM image shows two representative contacts to the top Ti electrodes as highlighted in yellow. It is the cross-conductance between such contacts that was used for this measurement. (b) The conductivities of SiNWs were measured throughout the Ti layer etching to ensure that SiNWs were not damaged. The SEM image (inset) shows the current pathway that was measured.

Using the SNAP technique, an array of 400 Pt NWs was then deposited over the Ti/SiO_2 layer and perpendicular to the underlying Si NWs. For the deposition of the Pt NWs, a different epoxy mixture (5 ml of THF, 5 drops of dibutylphthalate, 10 drops of epoxy and 1 drop of curing agent), compared to the one used for the Si NW generation, was used. A larger portion of epoxy in the new mixture enabled to hold the SNAP masters more firmly while the epoxy mixture was being cured on a hot plate and the GaAs masters were being wet-etched.

Especially, the usage of the new epoxy mixture was essential for the deposition of the top SNAP nanowire array because the surface of the substrate became relatively rough throughout many previous steps. For the similar reason, the new mixture was less vulnerable to undercut in the following BCl_3 plasma etching step. The prevention of the undercut was most challenging task in the project because it could arise from many factors correlating one another (recess depth, strength of cured epoxy, directionality of plasma etching etc.). Finally, careful BCl_3 plasma etching (10 sccm, 5 mTorr, 30 W) was used to transfer the Pt NW pattern to the underlying SiO_2/Ti film, thus forming Ti NW top electrodes.

The global SOG etching down to ~ 50 nm thickness was also critical for this top SNAP nanowire pattern transfer. In the devices that maintained a thick SOG layer, thus a deep recess over the active memory region, the epoxy was trapped in the recess to form its thick layer. The thick layer of epoxy was susceptible to undercut during the BCl_3 plasma etching and thus to have shorting problems in the top SNAP nanowires. This shorting problem is very fatal in a device performance because the yield of independent bits will decrease significantly. The etch endpoint was determined by monitoring the cross-conductance of the top Ti NWs. Complete transfer of the Pt NW pattern to the underlying Ti film was indicated by a fall in the cross-conductance to about 10 nS.

Note that the crossconductance does not go to zero since the Ti electrodes, while physically separated, are still electrically coupled through the crossbar junctions and the underlying Si NWs. The health of the underlying Si NWs throughout the Ti-etching steps was also monitored as shown in figure. In most cases, the devices that skipped the SOG planarization step lost the Si NW conductance completely before the crossconductance fell down to a value corresponding to the complete NW pattern transfer, indicating that Si NWs were damaged significantly during the BCl_3 plasma etching. Once BCl_3 plasma etching is done, the device is ready for testing. SEM images for final devices are presented in figure at different resolution.

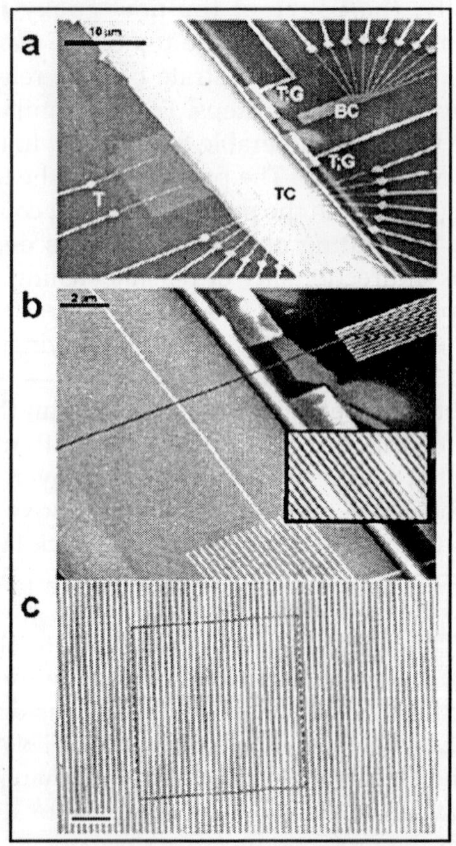

Fig. Scanning Electron Micrographs (SEMs) of the NW Crossbar Memory.

(a) Image of the entire circuit. The array of 400 bottom Si NWs is seen as the light grey rectangular patch extending diagonally up from bottom left. The top array of 400 Ti NWs is covered by the SNAP template of 400 Pt NWs, and extends diagonally down from top left. Testing contacts (T) are for monitoring the electrical properties of the Si NWs during the fabrication steps. Two of those contacts are also grounding contacts (G), and are used for grounding most of the Si NWs during the memory evaluation, writing, and reading steps.

Electron beam lithography patterned 18 top (TC) and 10 bottom (BC) contacts are also visible. The scale bar is 10 micrometers. (b) An SEM image showing the cross-point of top and bottom NW electrodes. Each cross point corresponds to an ebit in memory testing. (inset) The electron-beam-lithography defined contacts bridged 2-4 nanowires each. The scale bar is 2 micrometers. (c) High resolution SEM of approximately 2500 junctions out of a 160,000 junction nanowire crossbar circuit. The red square highlights an area of the memory that is equivalent to the number of bits that were tested. The scale bar is 200 nanometers.

160 KBIT MOLECULAR ELECTRONIC MEMORY CIRCUITS: DEVICE TESTING

The memory circuit was tested using a custom-built probe card and a Keithley 707A switching matrix for off-chip demultiplexing. Because SNAP NWs are patterned beyond the resolution of lithographic methods, each test electrode contacted between 2 and 4 NWs so that individual effective bit (ebit) contains between 4 and 16 crossbar junctions, but mostly 9 crossbar junctions. All ebits were electrically addressed within the 2D crosspoint array by the intersection of one Si NW bottom electrode and one Ti NW top electrode. Individual molecular junctions were set to their low resistance or "1" state through the application of a positive 1.5 – 2.3 V pulse (voltages are referenced to the bottom Si NW electrode) of 0.2 s duration. A junction was set to its "0" or high resistance state through application of a -1.5 V pulse, also of 0.2 s duration.

To avoid switching an entire column or row of bits, the switching voltage was split between the two electrodes defining the ebit. Thus, to write a "1" with +2 V, a single Si NW electrode is charged to +1 V, while a single Ti NW electrode is set to -1V, and only where they cross does the junction feel the full +2 V switching voltage. Half-selected bits, that is, bits receiving only half the switching voltage, were never observed to switch. This half-select issue, though being a clear drawback of crossbar architecture, is overcome by distinctive characteristic of rotaxane: As introduced, the voltages required to switch on/off MSTJs were uniform over broad junctions such that a half of the voltage did not perturb the junctions.

Individual ebits were read by applying a small, nonperturbing +0.2 V bias to the bottom Si NW electrode and grounding the top Ti NW electrode through a Stanford Research Systems SR-570 current pre-amplifier. Bits not being read were held at ground to reduce parasitic current through the crossbar array.

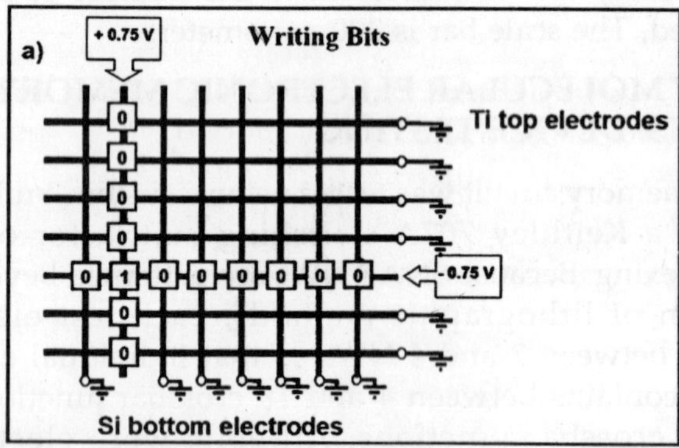

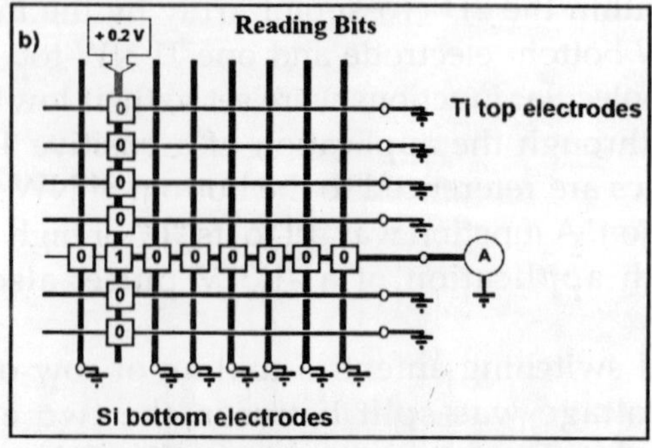

Fig. Writing and Reading Procedures in Crossbar Memory Measurements.

Note that all the electrical writing and reading operations described herein were done sequentially. Schematic illustrations describing the device testing procedures composed of writing and reading bits are presented in figure.

(a) Due to the half select issue, the writing bias was split into two halves of opposite polarity and each half was applied to both top and bottom electrodes, respectively, defining a designated cross-point. Other bits along these top and bottom electrodes are not perturbed due to the sharp switch-on/off bias characteristic of rotaxanes. (b) Before and after applying the writing voltages, the resistances of all bits are read at small non-perturbing reading bias to monitor the resistance change. Note that all other electrodes not involved in the switching of the designated cross junction stay grounded to minimize the parasitic current pathways.

By scanning electron microscopy inspection, the crossbar appeared to be structurally defect-free, with no evidence of broken, wandering, or electrically shorted NWs. Nevertheless, electrical testing identified a large number of defective bits and the nature of those defects. This testing was done by first applying a +1.5 V pulse relative to the Si NW bottom electrodes to set all bits to '1', and then reading each ebit sequentially using a non-perturbing +0.2 V bias. A -1.5 V pulse was then applied to set all bits to '0'. The status of each of the ebits was again read. The raw data throughout these procedures and the 1/0 current ratios are presented in figure.

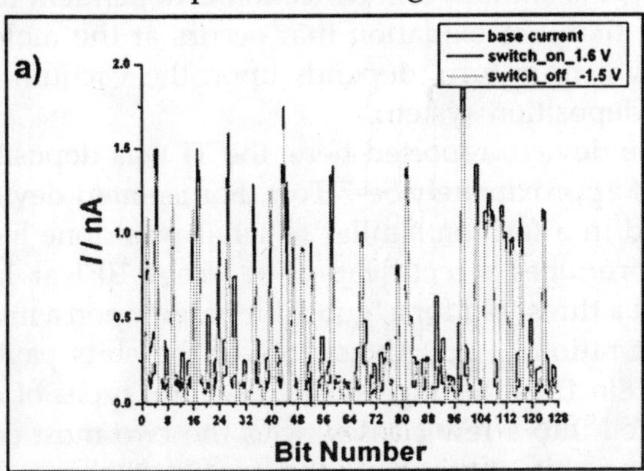

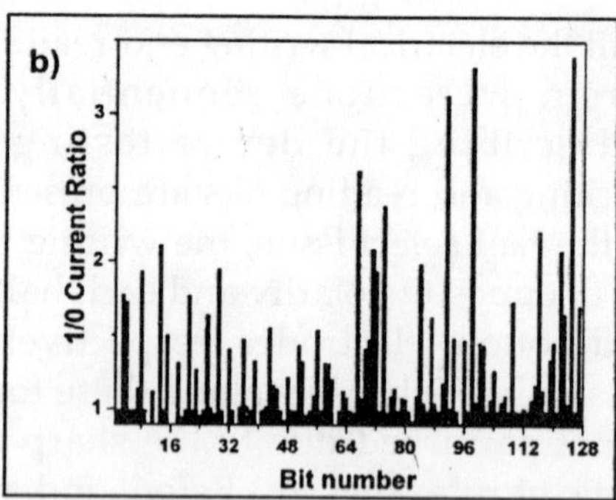

Fig. Data from evaluating the performance of the 128 ebits within the crossbar memory circuit. (a) raw current data when monitored at +0.2 V at the stage of before-switch on, after-switch-on and after-switch-off. (b) The current ratio of the '1' state divided by the '0' state of the tested ebits. Note that many of the ebits exhibit little to no switching response. Those ebits are defective.

About 50% of the bits yielded some sort of switching response. Some of that response, however, may have originated from parasitic current pathways through the crossbar array. This is an inherent drawback of crossbar architectures wherein each junction is electrically connected to every other junction. The standard remedy is to incorporate diodes at each crosspoint, and although the molecule/Ti interface yields some rectification, we additionally grounded all NW electrodes not being used during a read or write step. By the way, the amount of rectification is dependent upon the amount of titanium oxidation that occurs at the molecule/Ti interface which, in turn, depends upon the vacuum level of the metal deposition system.

For the devices reported here, the Ti was deposited at a pressure of approximately 5e-7 Torr. For isolated devices, but constructed in a fashion similar to what was done here, this typically produces a rectification of about 10:1 at 1 V. We established a threshold for a 'good' bit based upon a minimum 1/0 current ratio of ~1.5. About 25% of the ebits passed this threshold. Electrical testing revealed several types of defects. Bad ebits fell into a few classes, with the two most common groups being ebits that were either poor switches with little or no switching response or open circuits. Adjacent, shorted Ti top electrodes were identified when the ebits addressed by those electrodes were not independently addressable. Even though that type of defect is not completely fatal (i.e. two rows of fabricated ebits could still be utilized as a single row), we did not use ebits associated with shorted top electrode defects. The defects classified as 'switch defects' likely arose from sub-nanometer variations in the reactive ion etching process that was employed to define the top Ti crossbar NWs. Isolated devices, or crossbar memories patterned at substantially lower densities and with larger wires, can typically be prepared with a nearly 100% yield.

The switch defects led to only a proportional loss in the yield of functional bits, while bad contacts or shorted nanowires removed an entire row of bits from operation. An important result from the defect map is that the good and bad bits are randomly dispersed, implying that the crossbar junctions are operationally independent of one another.

	B1	B2	B3	B4	B5	B6	B7	B8
T1	1	2	3	4	5	6	7	8
T2	9	10	11	12	13	14	15	16
T3	17	18	19	20	21	22	23	24
T4	25	26	27	28	29	30	31	
T5	33	34	35	36	37	38	39	
T6	41	42	43	44	45	46	47	
T7	49	50	51	52	53	54	55	
T8	57	58	59	60	61	62	63	
T9	65	66	67	68	69	70	71	72
T10	73	74	75	76	77	78	79	80
T11	81	82	83	84	85	86	87	88
T12	89	90	91	92	93	94	95	96
T13	97	98	99	100	101	102	103	104
T14	105	106	107	108	109	110	111	112
T15	113	114	115	116	117	118	119	120
T16	121	122	123	124	125	126	127	128

Bad SINW contact
Good Switch
Poor Switch
Adjacent top NWs shorted (12.5%) (non-fatal

Fig. A map of the Defective and Useable ebits, along with a Pie-chart giving the Testing Statistics.

Note that, except for the bad Si NW contacts on bottom

electrodes B1 and B6, and the shorted top electrodes T2 and T3, the defective and good bits are randomly distributed. Type I defects (26% of the 128 tested) are ebits that exhibited an open-circuit conductance and a low or zero amplitude switching response when tested. Type II defects (22%) are non-switchable bits that exhibited a conductance similar to that of a closed bit.

However, the ultimate test of any memory is whether it can be used to store and retrieve information. Based upon the defect map, we identified the addresses of the usable ebits, and from those addresses configured an operational memory: the usuable bits were used to store and read out small strings of information written in standard ASCII code. The maximum number of ebits that could be tested was 180, but our electronics were configured to test 128 ebits (< 1% of the actual crossbar), and that was sufficient to demonstrate the key concepts of this memory.

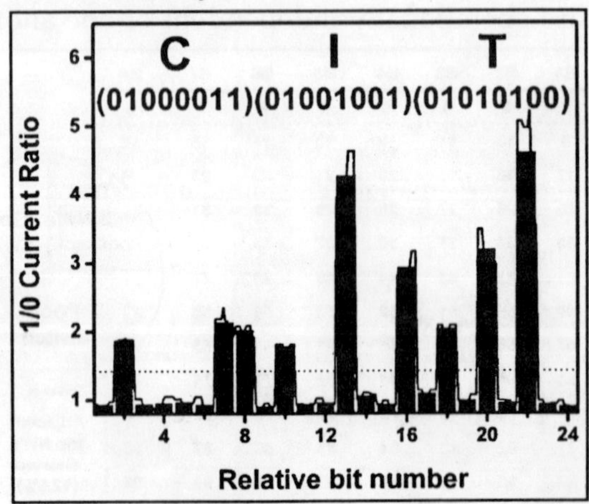

Fig. A demonstration of point-addressability within the crossbar. Good ebits were selected from the defect mapping of the tested portion of the crossbar.

A string of '0's and '1's corresponding to ASCII characters for 'CIT' (abbreviation for California Institute of Technology) were stored and read out sequentially. The dotted line indicates the separation between a '0' and '1' state of the individual ebits. The black trace is raw data showing ten sequential readings of each bit while the red bars represent the average of those ten readings. Note that deviations of individual readings from their average are well separated from the threshold 1/0 line.

The solid-state switching signature of the bistable rotaxanes that were used here has been shown to originate from electrochemically addressable, molecular mechanical switching for certain device structures, but not for metal wire / molecule/metal wire junctions. In fact, our desire to utilize molecular mechanical bistable switches as the storage elements is what dictated our choice of the silicon NW / molecule / Ti NW crossbar structure. This switching signature should be effectively size-invariant, meaning that it should scale to the macromolecular dimensions of these crossbar junctions.

Solid-state-based switching materials will likely not exhibit similar scaling since they arise from inherently bulk properties. The thermodynamic and kinetic parameters describing both the bistability and switching mechanism of the rotaxane switch (and similar molecular mechanical switches) have been quantified in a variety of environments. Those measurements required robust switching devices that could be cycled many times and at various temperatures.

The memory bits measured here were much more delicate – while all good ebits could be cycled multiple times, most ebits failed after a half-dozen or so cycles, and none lasted longer than ten cycles. However, we measured the rate of relaxation from the 1ÀÛÆÜ0 state for many of the ebits. From a device perspective, this represents the volatility, or memory retention time, of the bits.

With respect to the bistable rotaxane switching cycle, this represents a measurement of the rate limiting kinetic step within the switching cycle. Our measured rate (90±40 minutes; median decay = 75 minutes) was statistically equivalent to that reported for much larger (and more fully characterized) devices (58±5 minutes). Thus, our results are consistent with a molecular mechanism for the switching operation. The volatility measurements were carried out by switching selected bits to the "1" or low resistance state, and then reading the current through those bits as a function of time.

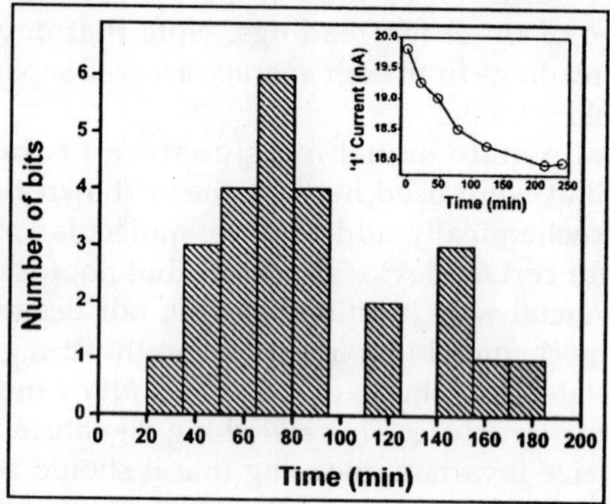

Fig. A Histogram Representing the 1/e Decay Time of the '1' State to the '0' State.

The 25 ebits represented in the data were each 'large' ebits, comprised of approximately 100 junctions, to increase the measurement signal to noise. Raw data from a single large ebit is shown in the inset.

LIMITATIONS OF THE SNAP PROCESS FOR CROSSBAR MEMORY

Formation

The nanofabrication methods described in this chapter for creating the 160 kilobit crossbar memory circuit can be significantly extended in terms of both memory size and bit density. For our memories, the crossbar electrode materials choices have proven to be very important for successful memory operation. In other words, Si bottom electrodes and metallic top electrodes with a Ti adhesion layer were keys. Metal NWs at 8 nm half-pitch have been reported previously.

Such NWs, formed by the SNAP process, only serve as templates for forming the crossbar electrodes. To be used in a crossbar memory, the SNAP NW pattern must be transferred to Si or Ti NWs for the bottom and top electrodes, respectively.

Thus, it is not just the SNAP process, but the ability to translate the initially deposited SNAP NWs to form other NWs that ultimately limits the size and density of the circuitry that can be fabricated. In figure, it is presented an array of 7 nm wide, 15 nm tall single crystal Si NWs patterned at 6.5 nm half-pitch. This corresponds to a crossbar that would contain approximately 6×1011 bits cm^{-2}.

While this array may not represent the density limit of what could be achieved, densities in excess of 1012 cm^{-2} will be very hard to obtain using these patterning methods. Similarly, the 160,000 junction crossbar also doesn't represent any sort of limitation. In figure we present SEM images of 1400 Si NWs formed using the SNAP method. Such an array size permits the formation of a 2 million bit crossbar, and it is certainly possible to further expand the concept to substantially larger structures.

As mentioned earlier, the primary limitation is that the SNAP process is that, while it is a parallel patterning method – since all nanowires within an array are created simultaneously, each array must be fabricated one at a time using a labour intensive process. A single worker, for example, can fabricate only about 20 arrays of Si NWs in a single day. However, recent advances in using nanoimprinting to replicate SNAP nanowires and to form crossbar structures indicate that high-throughput, parallel fabrication methods can be developed, even at the near molecular-densities described in this chapter.

This chapter focuses on molecular electronic memory circuits. The various generations of memory devices described in this chapter hold such common features that the devices are based on the crossbar circuit and utilize rotaxanes as information storage components. Through many generations, however, total number of bits within a single crossbar circuit and a bit density increased significantly. This scaling was possible due to the development of the fabrication procedures that allowed the integration of more delicate and higher density of electrodes with the rotaxane molecular monolayer.

Despite more complicated fabrication procedures, the

devices containing higher bit densities still showed the molecular switching signature. Especially, the final generation devices fabricated based upon the ultradense SNAP nanowire arrays also retained the molecular switching signature and exhibited a point addressability within a crossbar circuitry. Although about 75 % of the tested bits in the SNAP nanowire-based device turned out to be defective, the functional part was identified through an electrical testing and configured to write and read specific information. Notably, due to the extremely small pitch (~ 33 nm) of the SNAP nanowire array, the resultant 160 kbit crossbar memory circuits set a remarkable record in a bit density (10^{11} bits/cm^2).

Many scientific and engineering challenges, including device robustness, improved etching tools, and improved switching speed, remain to be addressed before this ultra-dense crossbar memory described here can be practical. Nevertheless, this 160,000 bit molecular memory does provide evidence that at least some of the most challenging scientific issues associated with integrating nanowires, molecular materials, and defect tolerant circuit architectures at extreme dimensions are solvable. While it is unlikely that these digital circuits will scale to a density that is only limited by the size of the molecular switches, it should be possible to significantly increase the bit density over what is described here.

Recent nanoimprinting results suggest that high-throughput manufacturing of these types of circuits may be possible. Finally, these results provide a compelling demonstration of many of the nanotechnology concepts that were introduced by the Teramac supercomputer several years ago, albeit using a circuit that contained a significantly higher fraction of defective components relative to the Teramac machine.

Chapter 4

Nanobiology Research

NANOBIOLOGY: FROM PHYSICS AND ENGINEERING TO BIOLOGY

Biological systems are inherently nano in scale. Unlike nanotechnology, nanobiology is characterized by the interplay between physics, materials science, synthetic organic chemistry, engineering and biology. Nanobiology is a new discipline, with the potential of revolutionizing medicine: it combines the tools, ideas and materials of nanoscience and biology; it addresses biological problems that can be studied and solved by nanotechnology; it devises ways to construct molecular devices using biomacromolecules; and it attempts to build molecular machines utilizing concepts seen in nature. Its ultimate aim is to be able to predictably manipulate these, tailoring them to specified needs.

Nanobiology targets biological systems and uses biomacromolecules. Hence, on the one hand, nanobiology is seemingly constrained in its scope as compared to general nanotechnology. Yet the amazing intricacy of biological systems, their complexity, and the richness of the shapes and properties provided by the biological polymers, enrich nanobiology. Targeting biological systems entails comprehension of how they work and the ability to use their components in design. From the physical standpoint, ultimately, if we are to understand biology we need to learn how to apply physical principles to figure out how these systems actually work. The goal of nanobiology is to assist in probing these systems at the appropriate length scale, heralding a new era in the biological, physical and chemical sciences.

Biology is increasingly asking quantitative questions. Quantitation is essential if we are to understand how the cell works, and the details of its regulation. The physical sciences provide tools and strategies to obtain accurate measurements and simulate the information to allow comprehension of the processes. Nanobiology is at the interface of the physical and the biological sciences. Biology offers to the physical sciences fascinating problems, sophisticated systems and a rich repertoire of shapes and materials. Inspection of the protein structure databank illustrates the breadth of scaffolds, shapes and properties that protein molecules and their building blocks can provide. Via a shape-guided self-assembly strategy, these can be put together toward a specific function. Further, by inserting synthetic non-natural residues at judiciously selected positions, or synthetic peptide linkers, we may selectively rigidify the construct, or obtain a totally new world of shapes and scaffolds. Such broadening of the chemical space may lead to an almost unlimited range of nanosystems and architectures.

Merging computation with experiment will accelerate nanodesign. Computational modeling will enhance the application of nanotechnology to key areas such as drug delivery and biomaterial design. Nanobiology is a field where interdisciplinary collaborations are essential and disciplines converge. Discipline convergence should enable the quantitation, leading to a better understanding of the regulatory networks within cells and between cells of an organism. These networks dictate how a cell responds to external stimuli, which in turn activate signaling cascades. It should allow the addressing of a broad range of questions on the structure and function of the cytoskeleton; the nuclear envelope; signal transduction by membrane embedded receptors; the nanomechanical properties of the extracellular matrix; nuclear transport; and voltage induced channel gating.

For successful nanostructure design, we need to figure out and be able to control the intermolecular associations. For a stable functional construct, there are two key elements: first, the conformations of the building blocks in the designed structure should follow their natural tendencies; and second, the associations should be favorable. Molecules interact through their surfaces. Thus, favorable associations derive from shape complementarity and contributions of the various physical components.

Nanobiology is in its infancy. Yet, biology provides an enormous range of engaging and stimulating problems with many *in vivo* examples of intricate, complex, fascinating biological systems. Understanding, mimicking and controlling the devices which target these processes and which are constructed from these molecules is a tremendous challenge to the converging disciplines in nanobiology.

NANOBIOLOGY

Nanobiology is an integrated area of nanoscience of polymer chemistry, pharmaceutical, biochemistry, bioengineering, biotechnology, molecular biology. The ultimate goal of nanotechnology is to discover and alternative choices with improvement in health science. The progress of nanotechnology allows us understanding how and what the relationship between the molecule and their consequent activity is combining with advanced instrument. Currently, we understand the function at molecular level, thus, the approach to achieve the goal of nanobiology is to mimic the function of macromolecule and applying the chemistry to synthesis the well-defined structure using biotechnology to identify the effectiveness of the system.

Current Status of Nanobiology

At present, scientists around the world are challenging the new nanomaterial in the forms of nanoliposomes, nanoemulsion, nanoparticles, nanomicelles, nanovesicles, nanofibers and nanoscaffold to satisfy the condition require for specific application in pharmaceutical and biomedical. A number of researches in this field are increasing drastically as evidence from the articles, patents and products on yearly basis.

Centre of Innovative Nanotechnology

Chulalongkorn University considers nanotechnology as a promising integrated knowledge in science, engineering, pharmaceutical and medical. On this view point, Centre of Innovation (CIN) was established on the occasion of 100th anniversary of the university. Nanobiology is a research cluster in CIN aiming for ultimate achievement in developing nanomaterial for health science. The cluster consists of professors, researchers, Ph.D. students and master students who are working on the nanoscience area. The cluster also participate nanotechnology master and Ph.D. programme by having teaching courses, organizing short courses and supporting student researches. At the present, nanobiology cluster is running on nanodrug delivery, drug targeting, nanoallergen delivery, nanoadjuvent including micelle and liposome projects.

Project in Progress and Future Plans of CIN on Nanobiology

Each year, nanobiology has its own research projects either the continuous or the short term projects on 2008. There are 4 projects related to microemulsion, liposome, and nanoparticles. As the projects not only deal with the material development but also the functions of the material both in vitro and in vivo, nanobiology cluster is planning to extend their researches by collaborating with pharmaceutical and medical school in Chulalongkorn University. The future plans also include the collaborative research with government and private sector especially R&D laboratories on novel drug development.

When most of us think about nanotechnology, we imagine high-tech, ultra-fast computer chips, new stain-resistant materials, or even fictional, self-replicating nanomachines. While these marvels of nanotechnology are usually associated with the technological exploits of physics, chemistry and engineering, a new type of nanoscience is being explored in laboratories across the world. The exciting new field of nanobiology has taken centre stage at the interface between two worlds, the physical and the biological. The biological world that most of us experience is typically on the "macro" scale. The plants, animals, and other humans that we interact with are usually centimeters to meters in size and can be seen with the naked eye. When we move down to the cellular level, we start seeing cells on the order of one to tens of micrometers (one-millionth of a meter). Stepping down yet another size scale, biological components such as DNA and cell membranes are on the order of 2-3 nanometers (one-billionth of a meter) while proteins, such as antibodies, are 5-10 nanometers in size. Since all living things share these common components (DNA, proteins, and membranes), biology is, and always has been, living at the nanoscale.

Nanobiology, as a field of study, signifies the merger of biological research with nanotechnologies such as nanodevices, nanoparticles, or unique nanoscale phenomena. Although molecular biologists have been working with nano-sized biomolecules for the last few decades, nanobiology was not defined as a discipline until researchers started making a focused effort to use our knowledge of nanotechnology to tackle biological problems. As a merger between nanotechnology and biology, nanobiology encompasses a wide range of research topics that can be divided into two basic categories:

- Nanotechnologies applied to biological systems,
- The development of biologically-inspired nanotechnologies.

One reason to split nanobiology into these categories is to differentiate between the sources of inspiration for the research. In the first category, we use our physical, chemical and engineering knowledge to better understand biology, visualize and detect biological processes, or to create better ways of interfacing the biological and physical worlds. This technical approach towards biology relies upon our abilities to imagine and create systems that can be used for biological research. On the other hand, biologically-inspired nanotechnologies use biological systems as the inspiration for technologies that we seek to create. This veritable "look in the rearview mirror" allows us to learn from eons of evolution that have resulted in highly elegant, naturally created systems. This type of nanobiological research can be summarized as a form of "biomimetics", which seeks to "mimic" biological systems or biological structures. While these two main categories of nanobiology differ in their general approach, they can both be used to better understand the crossover between our biological and physical worlds.

To better understand nanobiology as a field of study, it is helpful to look at some of the general research topics that are being studied both in academia and commercial settings. In very general terms, these areas of study can be divided into nanobiological structures and systems, biomimetics, nanomedicine, nanoscale biology, and nano-interfacial biology. By grouping research into these fields of study, we can observe general ways in which nanotechnology and biology are brought together for common research goals.

NANOBIOLOGICAL STRUCTURES & SYSTEMS

Nanobiological structures and systems research can include a wide variety of technologies and biological systems, but mainly focuses on using nanotechnology to detect, measure, or probe biological systems.

The advantage of using nanotechnology for these purposes comes from the unique physical properties that can be achieved at the nanoscale. For instance, nanotechnology can be used to create nanochips or nanopatterned devices to screen large numbers of biological targets. Because of the small size of these systems, researchers can use smaller sample sizes, perform faster analyses, or use smaller amounts of expensive chemicals and reagents.

In addition, unique physical phenomena at the nanoscale can be harnessed for sensing, detection, and analytical purposes. Many electrical and optical properties that occur at the nanoscale are responsive to biological molecules, yielding highly sensitive analytical techniques. The topics listed below represent some of the most relevant research areas within nanobiological structures and systems, and some of the most high profile research that is going on at major research centers, including the College of Nanoscale Science and Engineering (CNSE) at the University at Albany.

- Lab-on-a-chip systems and sensors
  (low-power, portable sensors that incorporate multiple analytical steps into one system)
- Biosensors
  (sensors that can detect biological molecules, cells, or biological processes)
- High throughput / massively parallel sensors
  (sensors that detect many targets at the same time or in a rapid manner)
- Ultra-small sample volume sensors
  (sensors that use very small volumes of chemicals or reagents)
- BioMEMS (Biological Micro Electrical Mechanical Systems)
  (micrometer-sized mechanical and electrical "machines" coupled with biological molecules or cells)

BIOMIMETICS

Biomimetics, or the study of biological systems to inspire human-engineered systems, is a unique area of study that relies on natural systems for design concepts. The number of possible research areas within biomimetics is large, since this field seeks to broadly inspire technological and engineering advances from biological themes. A general example of biomimetics is the use of the lotus plant as inspiration for water-repellent technology.

At the nanoscale, the leaves of the lotus plant have regularly-spaced features that cause water droplets to roll off of the leaf surface, without spreading out and "wetting" the leaf. This natural water-repellency has been mimicked by constructing similar nanostructures out of polymeric materials. These bio-inspired nanostructures accurately mimic the properties of the lotus leaf, creating a non-wettable, water-repellent surface. Other research areas that fall under the theme of biomimetics are listed below:

- Bio-inspired architecture for nanotechnology (from cells, viruses, proteins, and other biomolecules)
- Chemical/structural mimicking of biology (for sensors and analytical systems)
- Animal-on-a-chip and mock organs/systems (for drug testing, drug delivery, and simulating environments)

NANOMEDICINE

Nanomedicine is a broad topic area that can encompass many of the other research areas within nanobiology. For our purposes, however, we can define nanomedicine as the application of nanotechnology towards the medical field. This can include the development of new types of sensors and analytical tools, as well as nanoscale methods of delivering therapeutic drugs or diagnosing disease. Many of the exciting advances within nanotechnology are beginning to be harnessed by the medical field.

Nanoparticles and nano-engineered substances have been used for drug delivery and similar systems have been used to target disease-causing agents and tumor cells for therapeutics. In addition, nanoparticles, such as fluorescent quantum dots are beginning to be explored for advanced imaging and diagnostics. A brief list of nanomedical research topics is listed below:

- Nano drug delivery and therapeutics
- Nanodevices for imaging, sensing, and analytical purposes
- Nanoparticle and quantum dot labeling for diagnostics
- Nanoparticle-based therapies for disease

NANOSCALE BIOLOGY

Nanoscale biology is a general description of basic biological research that is either performed on the nanoscale, or that is aided by nanoscale technologies. This encompasses multiple research topics that are often hard to group within some of the more common nanobiological research categories.

At the molecular level, all biological systems are made up of nanoscale components. The DNA, RNA, lipids, carbohydrates and proteins that make up each of our cells are all nanoscale molecules that can be studied or manipulated using nanotechnology.

One reason to study biology at the nanoscale is to observe properties that may not be seen at the micro and macro size scales. For instance, measuring the physical properties of individual proteins or DNA molecules could give us additional insight into their structure and function. This knowledge can be used to better understand how biological systems operate and how different components of biological systems work together to make living things move, grow, interact, or even reproduce. Just about any biological system can be approached from the nanoscale, yielding new information that cannot be observed at other size scales.

- Cellular-level studies (electrical, optical, force measurements with nano-tools)
- Molecular-level studies (DNA, RNA, lipid, carbohydrate, & proteins)
- Utilizing nanoscale tools for unique biological studies (to better understanding protein folding, DNA replication, etc.)

NANO-INTERFACIAL BIOLOGY

Nano-interfacial biology brings together chemistry, biochemistry, materials science, and nanotechnology. In many ways, research efforts in nanobiology (and other nanosciences) are greatly dependent upon chemical reactions and interactions at surfaces. This is especially important in nanobiology since most biology occurs either near physical surfaces, or near the surfaces of proteins, enzymes or membranes. Therefore, this research field focuses on chemical and biochemical interactions with interfaces at the nanoscale.

One interesting component of nano-interfacial biology is the "self-assembly" of biological molecules. Many biological molecules can form larger, defined structures without needing any blueprints, plans, or complex nano-construction crews.

This self-assembly process creates larger, three-dimensional structures that often have regularly repeating patterns at the micro and nanoscale. Self-assembly is therefore an attractive method for constructing nanoscale structures, since it greatly simplifies the fabrication process. A major push in nanobiology is to combine physical interfaces with self-assembling molecules to create unique nanoscale structures or devices.

By optimizing how biological molecules or cells interact with these physical surfaces, we can begin to develop hybrid systems that are biologically powered, biologically actuated, or even nanoscale devices that trigger biological responses.

This hybrid approach could greatly benefit prosthetic or implanted devices that rely on the seamless interaction between an organism and the prosthetic/implanted device. One could even imagine interfacing complex electrical systems, like microchips, with an organism's nervous system or brain.

For these types of systems, some of the most important interactions are those that occur at the physical-biological interface. Some relevant research areas in this field are listed below:

- Self-assembly and patterning of biomolecules on physical surfaces
- Immobilization of proteins, enzymes, and nucleic acids at physical interfaces
- Molecular interactions at surfaces and nanostructures
- Surface chemistry including self-assembled monolayers, biochemical surface attachment, chemical and biochemical surface patterning
- Development of molecular and cellular interfaces for prosthetics & implanted devices

From this survey of research topics within nanobiology, it should be clear that nanobiology encompasses a vast array of biological, chemical, physical and engineering research. Unlike traditional biological studies, nanobiology focuses on using our understanding of nanotechnology to advance our research capabilities and approach biology from a unique, nanoscale perspective. Similar to physics and materials research, many systems have unique properties at the macro, micro and nanoscale levels.

By approaching biology from the nanoscale, or by approaching nanotechnology from a biological perspective, we hope to gain unique insights into how biological systems function and how we can develop new and better bio-inspired technologies. This exciting frontier promises to develop new techniques, provide new understanding, and as in any scientific field, yield new and ever more complicated questions to be answered.

GETTING STARTED

Amazing things happen at the boundary between quantum physics and the 3D world of Netwonian physics. Like life, for example. I've been reading a lot about recent research in molecular biology and the basic building blocks of life at the molecular nanoscopic level. The following are the notes I've been making along the way to consolidate my understanding, including quotes and pictures from a lot of very cool web sites and links to multimedia presentations that vividly show how amazing the world is at the nanometer scale. Hope we find them interesting, fun and thought provoking.

Water

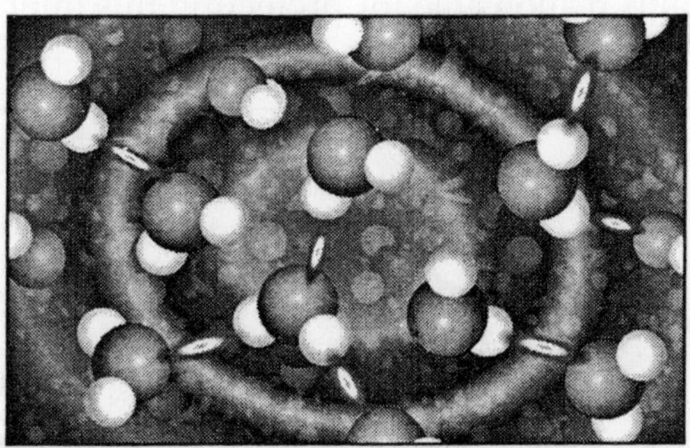

Apparently, *water* is the 'key ingredient' that makes all of the complex cellular mechanics of life possible. It has some very unique 3D characteristics that make it possible for cells to manipulate molecules to construct a wide variety of different compounds out of a small number of building blocks - e.g. to make a huge variety of different protein molecules out of a set of about 20 amino acids.

Water

Nature shows that molecules can serve as machines because living things work by means of such machinery. Enzymes are molecular machines that make, break, and rearrange the bonds holding other molecules together. Muscles are driven by molecular machines that haul fibers past one another. DNA serves as a data-storage system, transmitting digital instructions to molecular machines, the ribosomes, that manufacture protein molecules. And these protein molecules, in turn, make up most of the molecular machinery.

Life's local data storage is, of course, the DNA strands, broken into specific genes on the chromosomes. The task of instruction-masking (blocking genes that do not contribute to a particular cell type) is controlled by the short RNA molecules and peptides that govern gene expression.

The internal environment the ribosome is able to function in is the particular chemical environment maintained inside the cell, which includes a particular acid-alkaline equilibrium (pH between 6.8 and 7.1 in human cells) and other chemical balances needed for the delicate operations of the ribosome. The cell wall is responsible for protecting this internal cellular environment from disturbance by the outside world.

In a liquid state, the two hydrogen atoms make a 104.5° angle with the oxygen atom, which increases to 109.5° when water freezes. This is why water molecules are more spread out in the form of ice, providing it with a lower density than liquid water. This is why ice floats.

Although the overall water molecule is electrically neutral, the placement of the electrons creates polarization effects. The side with the hydrogen atoms is relatively positive in electrical charge, whereas the oxygen side is slightly negative.

So water molecules do not exist in isolation, rather they combine with one another in small groups to assume, typically, pentagonal or hexagonal networks. The partially positive hydrogen atom on one molecule is attracted to the partially negative oxygen on a neighboring molecule (hydrogen bonding).

Three-dimensional hexamers involving 6 molecules are thought to be particularly stable, though none of these clusters lasts longer than a few picoseconds; they can change back and forth between hexagonal and pentagonal configurations 100 billion times a second. At room temperature, only about 3 percent of the clusters are hexagonal, but this increases to 100 percent as the water gets colder. This is why snowflakes are hexagonal.

These three-dimensional electrical properties of water are quite powerful and can break apart the strong chemical bonds of other compounds. Consider what happens when we put salt into water. Salt is quite stable when dry, but is quickly torn apart into its ionic components when placed in water. The negatively charged oxygen side of the water molecules attracts positively charged sodium ions (Na^+), while the positively charged hydrogen side of the water molecules attracts the negatively charged chlorine ions (Cl^-).

In the dry form of salt, the sodium and chlorine atoms are tightly bound together, but these bonds are easily broken by the electrical charge of the water molecules. Water is considered "the universal solvent" and is involved in most of the biochemical pathways in our bodies. So we can regard the chemistry of life on our planet primarily as water chemistry.

MEMBRANES - KEEPING THE INSIDE IN AND THE OUTSIDE OUT

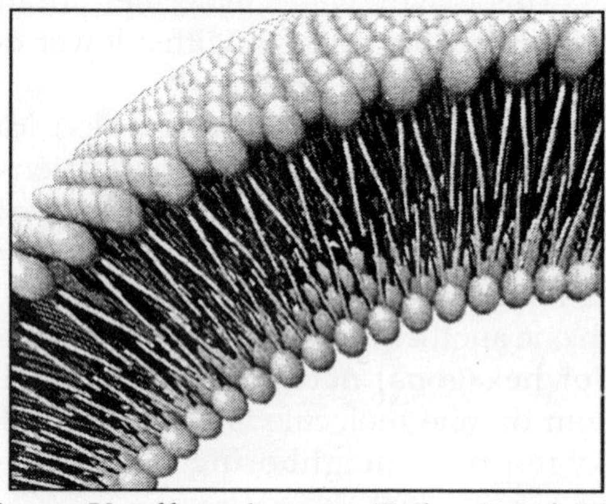

From Stuart Kauffman's essay "What is Life?" in the book "The next fifty years": An enclosed gas in a thermodynamically isolated box can do no work. But if the box is divided into two parts by a membrane, then one part can do work on the other part; for example, if the gas pressure is higher in the first part, the membrane will bulge into the second part, doing mechanical work on it.

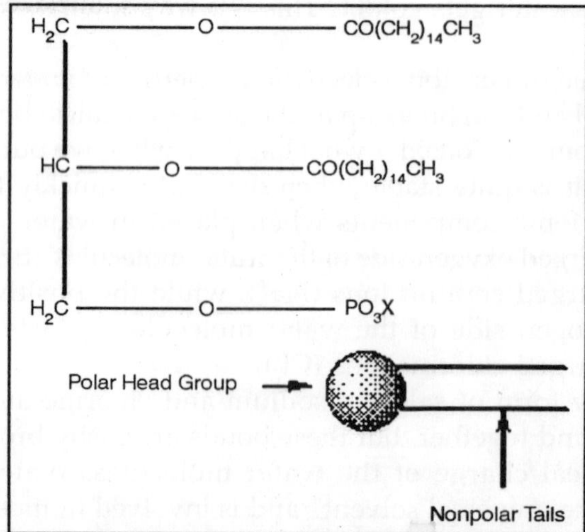

Thus work cannot be achieved in the universe unless the universe is divided into at least two regions. Furthermore, just where did the membrane come from? For cells, the membrane is made up of molecules that are polarized on one end - i.e. their atoms are physically arranged so that another molecule can get close enough to the molecule to feel the inter molecular force exerted by the presence or lack of electrons around one atom - and unpolarized at the other end. The polarized ends are 'hydro-philic' and are attracted to water, whereas the non-polar ends are hydrophobic and get pushed away from the water molecules.

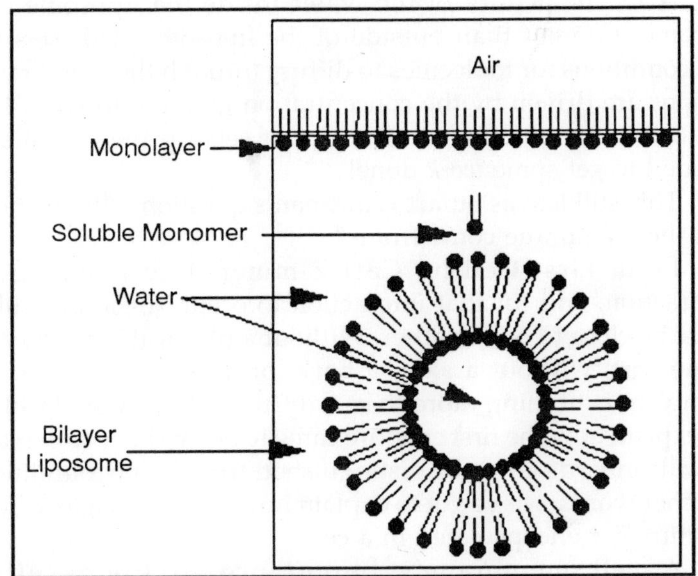

How would we orient this molecule in water? There are several possible ways. A small number of these molecules might be soluble in water as above. Mostly, however, the nonpolar tails wants to get out of the water, while the polar head like to stay in the water. Again, some of the molecules migrate to the surface of the water, with the nonpolar tails sticking out into air, away from water, to form a monolayer on the top of the water.

Others will self- aggregate, through Inter-Molecular Forces (IMF's) to form a bilayer or membrane. Because there are two tails per head group, the tails can't pack together as tightly. Imagine the bilayer or membrane curving around and eventually meeting. A structure like this would look like a small biological cell. The interior of this little cell, or liposome, is filled with water.

Now, cell membranes are not rigid or impermeable. In fact, things like proteins with one hydrophobic end can push into them and get embedded in them fairly easily. Ions (like Na^+ or K^+, for example) are not able to pass through the membrane.

Also, the acidity of the water inside the liposome can become different than outside of the liposome. This sets up the conditions for molecules to diffuse through the membrane, as ions are driven by the concentration gradient to spread to regions of lower concentration. i.e. it sets up the conditions needed to get some *work* done!

This still leaves Stuart Kauffman's question - "Just where did the membrane come from?"

From First Cell by Carl Zimmer: Life is chemical interaction, and for that interaction to occur, life's molecules must be close to one another. Without a physical boundary of some sort, without a skin, a bark, or a cell membrane, an organism is nothing more than a diffusing blur of molecules. To explain how the first creature came to be, we have to explain how its innards got to be distinguished from its surroundings. In other words, we've got to explain how the first single- celled creature got encapsulated in a cell.

A cell membrane's importance to life is often underappreciated, says David Deamer. People say, 'Well, it's just a little bag.' But it's much more. It's the interface between life and everything that's outside. The membrane of any cell has to do many things at once. It has to be impermeable enough to keep essential things (like DNA) in and harmful things (like viruses and poisons) out. Yet a cell membrane can't form a perfect seal. It has to be able to flush out waste and heat from its own system and take in nutrients from the surrounding medium. And the first cell membrane, like the membranes of many single-celled organisms today, probably had to be able to collect energy as well.

FUN WITH MOLECULAR ORIGAMI

When we start reading about how the brain works and how cells work, we come across a number of key words over and over. Words like 'ion channel', 'neurotransmitter', 'receptor'. Well, what the heck are these things, really? Finding the answer to that question has dragged me through a heck of a lot of interesting territory. The short answer is 'protein'. They are all made of protein. So, before diving into receptors and stuff, here's some info I've found on proteins:

Biochemistry toolkit

From all of that, we know that proteins are built out of a 'toolkit' of 20 amino acids. An amino acid looks like this:

i.e. a central carbon attached to a carboxylate (the COOH), an amino (the NH_2) and a hydrogen atom. The differences between the 20 types are in the side groups that can be designated as R. R can either be hydrophobic or hydrophilic.

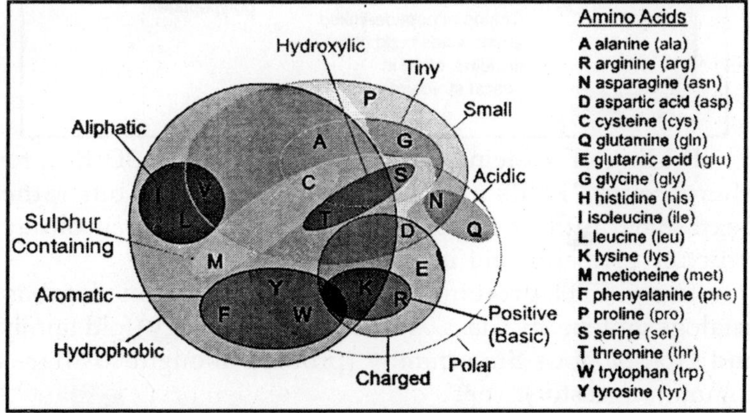

Protein is made by stringing together a bunch of these amino acids with peptide bonds.

The body builds around 25000 different types of proteins, some of which act as nano-scale engines (e.g. in muscles), others containing 500-1000 amino acids act as enzymes, some are used as messengers within the cell or between cells, and still others that have both a hydrophobic amino acid side group and a hydrophilic amino acid side group embed themselves in the cell wall and form either ion channels or receptors. The 3D shape of the protein controls how other molecules interact with the protein.

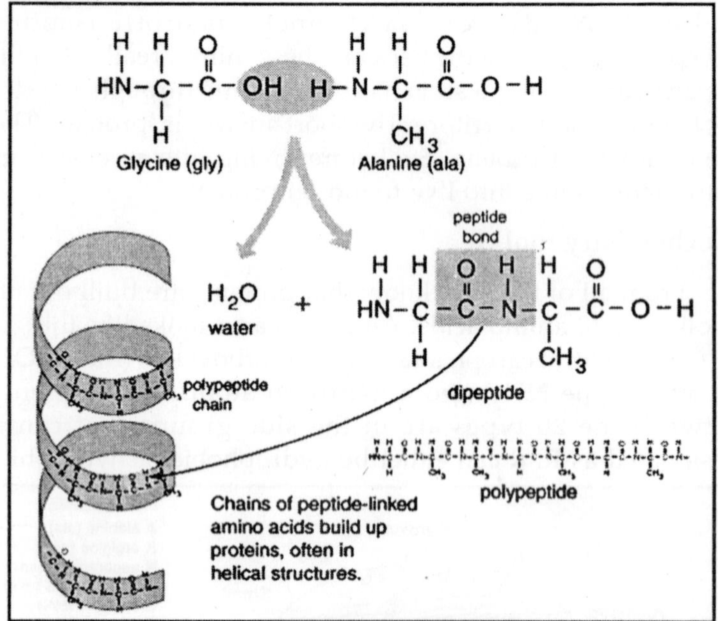

Folding of proteins can be examined using 3D Electron Microscopy. "Folds are not "fixed" structures but rather flexible definitions related with the number of secondary structure elements and its spatial distribution."

The CATH protein structure classification database catalogs proteins by Class, Architecture, Topology (fold family) and Homologous Superfamily (proteins thought to share a common ancestor).

Unraveling the Mystery of Protein Folding provides a fascinating overview of recent discoveries about the 3D mechanics of proteins, including the role that improperly folded proteins have been found to play in Alzheimer's disease. IBM's Blue Gene supercomputer is one of the most powerful tools being used in this area of study. It allows researchers to model the 3D forces at play that result in protein folding.

Protein formation: Codones, Histones and Ribosomes

We've come a long way...

1831: Cell nucleus first described by Robert Brown

1871: DNA of cells first isolated by Frederich Miescher in pus of wounds

1938: Ribosomes, which means "body of ribose" after these organelles were found to contain the sugar ribose, were first described by Albert Claude and were observed in animal cells which had been infected with Rous sarcoma virus. He then ascertained that noninfected cells also contained these particles and first called them microsomes (which in part they were, because ribosomes are bound to microsomes).

1943: Ribosomes in bacterial cells first described by Luria, Delbruck and Anderson.

1950: George Palade and Keith Porter using Electron Microscopy describe Ribosomes when they were at Rockefeller Institute, and noticed their location on the endoplasmic reticulum and free in cytoplasm.

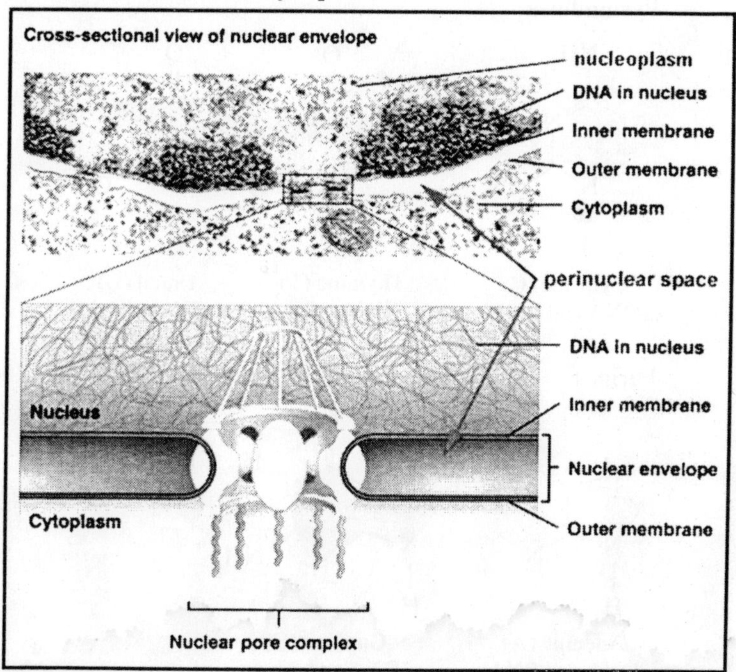

1953: Watson and Crick describe the double-helix structure of DNA.

2003: Completion of sequencing the human genome.

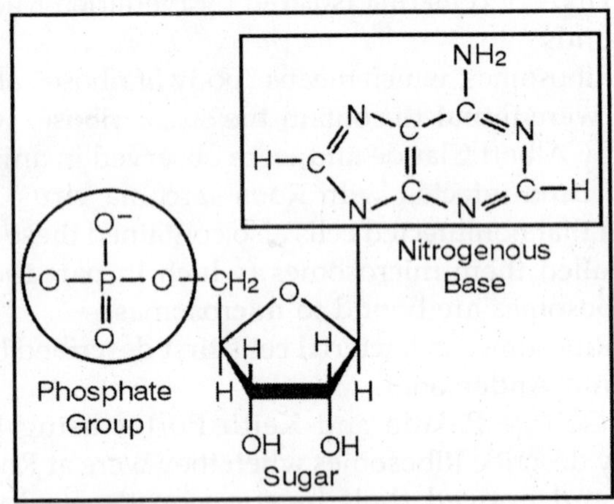

Pyrimidines

Cytosine (C)
(DNA and RNA)

Thymine (T)
(DNA only)

Uracil (U)
(RNA only)

Purines

Adenine (A)
(DNA and RNA)

Guanine (G)
(DNA and RNA)

DNA or deoxyribonucleic acid is a large molecule structured from chains of repeating units of the sugar deoxyribose and phosphate linked to four different bases: Adenine, Thymine, Guanine, and Cytosine. Ribonucleic acid (RNA) is a single-stranded copy of DNA, in which the Thymine is replaced by Uricil. During mitosis (cell division), each DNA molecule is wrapped around histones and wrapped further up into chromatin. During interphase (between divisions), chromatin is more extended, a form used for expression genetic information. The DNA is loosely held in the nucleus. As well, this protein synthesis movie is a cool little that provides a bit more detail on how protein is synthesized, including the role of enzymes during a step called acylation, where tRNA molecules are linked to their respective amino acids, forming complexes called aminoacyl-tRNAs.

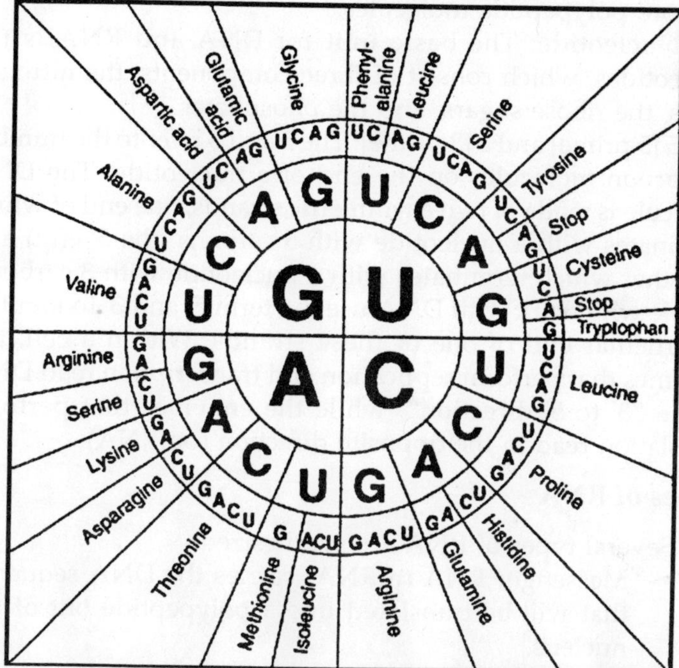

DNA encodes the unique nature of different organisms by specifying the precise structure of each protein in a cell. In analogy to DNA, proteins are made from a linear sequence of amino acids, and the exact sequence of amino acids is what determines the function of the protein. A DNA sequence is translated into the protein sequence by a code, where a triplet of bases specifies a single amino acid; some codons specify the end of the protein.

As a cell "reads" the DNA instructions, it builds a protein by adding successive amino acids one at a time, as defined by the codons. With 64 possible codons (4 DNA bases in the 3 positions of the codon, or 43) and only 20 standard amino acids, some amino acids can be specified by more than one codon. Some key words:

Transcription = DNA '! RNA

Translation = RNA '! protein

Polypeptide = a molecule made of a sequence of amino acids linked by peptide bonds. Proteins are made up of one or more polypeptide molecules.

Nucleotide: The basic unit for DNA and RNA is the nucleotides, which consist of three components: the nitrogen bases, the ribose sugars, and the phosphates.

5'[5 prime] and 3'[3 prime]: The 5 and 3 refer to the number of Carbon molecules on the end of a nucleotide. The DNA molecule is made of two asymmetric strands, one end of which terminates with a nucleotide with 5 carbons (the 5' end) and the other which terminates with a nucleotide with 3 carbons. People who work with DNA use the term 5' and 3' to identify a particular end of one of these strands. Within a cell, the enzymes that perform replication and transcription read DNA in the "3' to 5' direction", while the enzymes that perform translation read in the opposite direction (on RNA).

Types of RNA

Several types of RNA are synthesized:

- Messenger RNA (mRNA) carries the DNA sequence that will be translated into a polypeptide out of the nucleus.
- Ribosomal RNA (rRNA) is used in the building of ribosomes
- Transfer RNA (tRNA) carry amino acids to the

ribosomes to add to the polypeptide chain that is being constructed.

- Small nuclear RNA (snRNA) mediate the processing of the genes that are transcribed to produce the large precursor molecules ("primary transcripts") that are used to build mRNA, rRNA, and tRNA. Unlike most polypeptides, construction of these molecules must be done inside of the nucleus.
- Small nucleolar RNA (snoRNA) are used within the nucleus for several different functions.
- MicroRNA (miRNA) are tiny RNA molecules that "appear to regulate the expression of messenger RNA (mRNA) molecules."
- XIST RNA inactivates one of the two X chromosomes in female vertebrates.

Transfer RNA (tRNA)

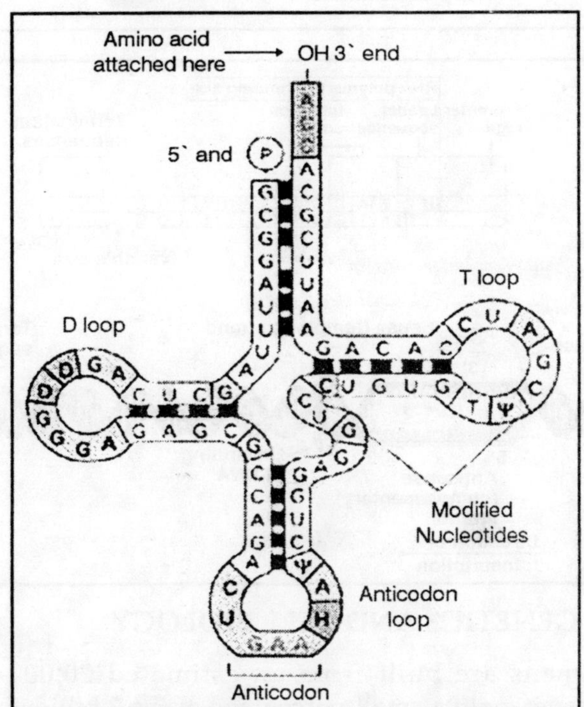

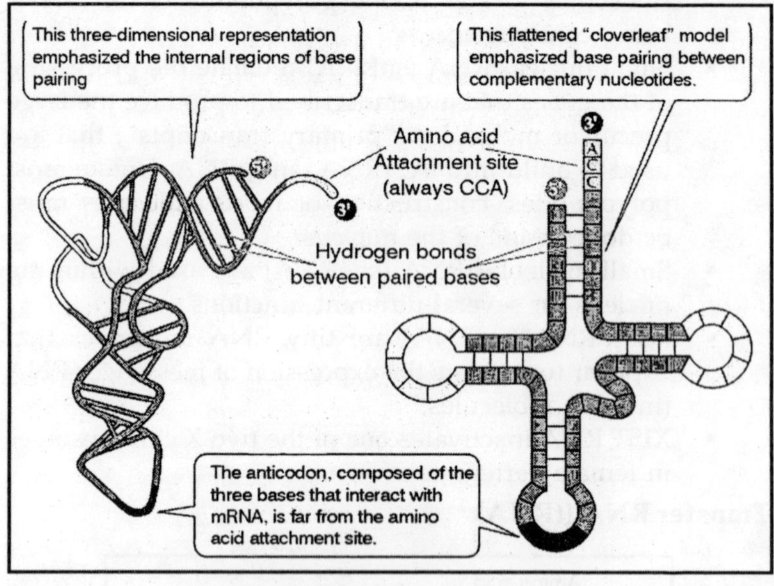

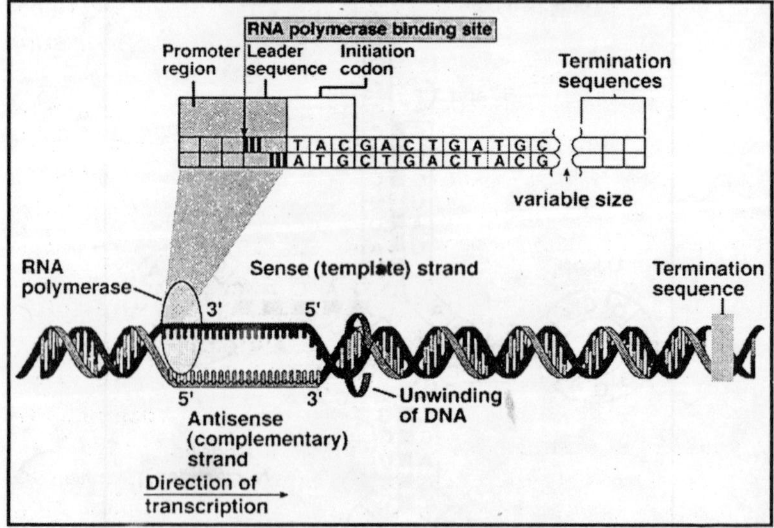

BASIC GENETICS AND CELL BIOLOGY

Humans are built from an estimated 20,000-35,000 proteins, yet, only a small percentage of the 3 billion bases in the human genome codes for these proteins. The discrete sections of DNA that encode proteins are referred to as "genes." A gene contains special codons that tell the cell where the gene - and hence, the protein - start and stop. Between these signals, the regions of the DNA that code for amino acids (the exons) are separated by a variable number and length of non-coding DNA sequence (the intervening sequence, or introns).

Protein production is a highly regulated process. For example, a cell does not want to waste energy making the proteins needed for cell division if it is busy with other functions, such as secreting a hormone.

This process of turning a gene on and off depending on the cell's need for a particular set of proteins is referred to as regulation of gene expression. Certain proteins, and even other regions of DNA, physically bind to the DNA sequence surrounding a gene to affect its expression. This interaction occurs at regions of the DNA with apt names, such as promoters or enhancers of gene expression.

Gene regulation is an essential part of life. Since every cell in an organism contains the same genetic blueprint, different cell types are created by turning on different genes at different times during development. In fact, it is differential gene expression that allows stem cells to become unique cell types. Gene regulation is also critical for cellular response to metabolic needs.

There are a number of different ways that are used to control which genes are activated (and thus which proteins, etc. are generated), and the rate at which the resulting molecules that are created at. The most important and widely used mechanism is by altering the rate of transcription of the gene.

Alternative mechanisms include:

- Changing the rate that RNA transcripts are processed within the nucleus
- Reducing the stability of mRNA molecules that are created, so that they degrade more quickly
- Changing the efficiency that ribosomes are able to translate the mRNA into a molecule

More about Histones

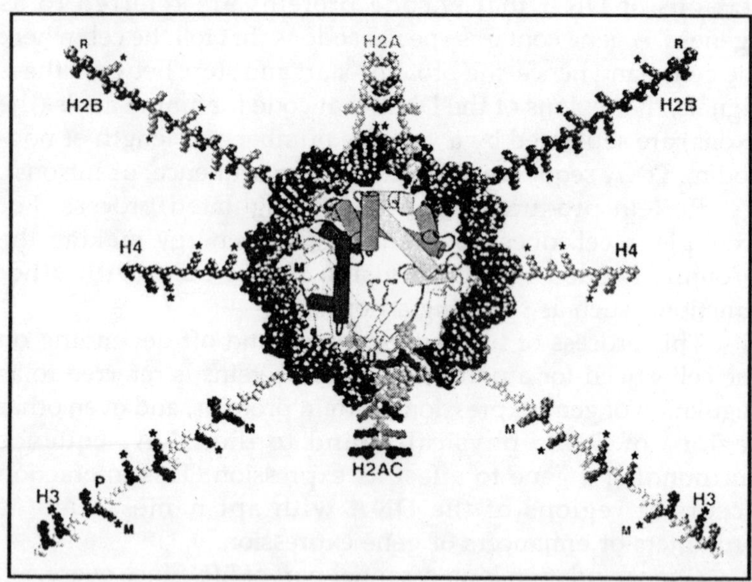

The preceding blog entry touched on how the DNA molecule is wound around histones, then further wound into a fibre of chromatin which is then finally wound up to form chromosomes. The way DNA winds around histones is something that is undergoing active research, because it looks like it may play an important role in regulating transcription.

Scientists believe that histone modifications are crucial actors in the activation and repression of gene expression. Histones help to package DNA, the hereditary material of life, into each cell's nucleus. The double-helical strand of DNA wraps around a ball of histones consisting of four distinct proteins: H2A, H2B, H3 and H4. The genes that encode the H3 and H4 proteins used in histones are so fundamental to the actual storage of the genome that they have a zero effective mutation rate. Almost the exact same genes are found in humans, chickens, grass, and mould.

In 1997, Roeder and Wei Gu, then a postdoc in Roeder's lab, showed that an enzyme called p300/CBP, known to modify the chemical composition of histone tails, serves as a transcriptional coactivator for p53. Coactivators are regulatory proteins that, together with activator proteins, are required to turn genes on.

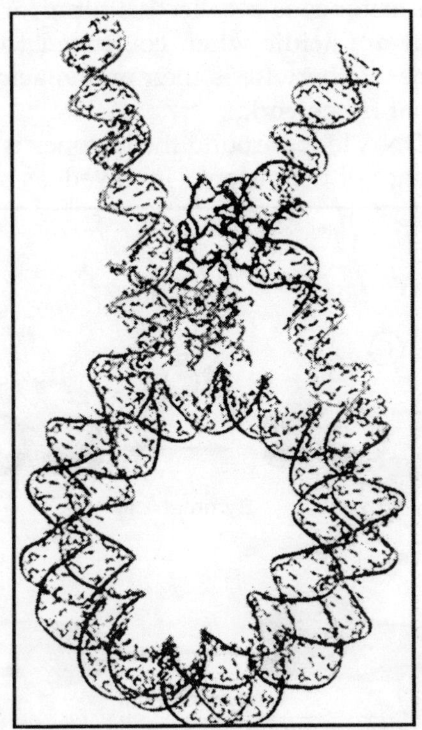

In 2002, An, Roeder and their Rockefeller colleagues provided important new information about the role of histone tails in gene activation. Scientists knew that histone tails repressed gene activation by preventing transcription factors — proteins that help read out the information encoded in DNA — from gaining access to DNA. An and colleagues, using a test tube system of coiled up chromatin created from engineered or recombinant histones and DNA, showed that — as Allfrey and Allis had predicted — histone tails and associated modifications are required for reversing the repression of transcription and that p300 plays an important role in this process as a histone-modifying enzyme.

Until a few years ago was it generally assumed that the

histones take part in the selective transcription of single DNA segments (genes or groups of genes). This proved to be wrong: a group of acidic nuclear proteins (non-histones) fulfils this function. Non-histones is a collective term for a number of regulators that are acidic when compared to the strongly alkaline histones. Otherwise is their amino acid composition within the usual framework.

The way DNA loops around the histones is controlled by the characteristics of the histones involved, as shown below:

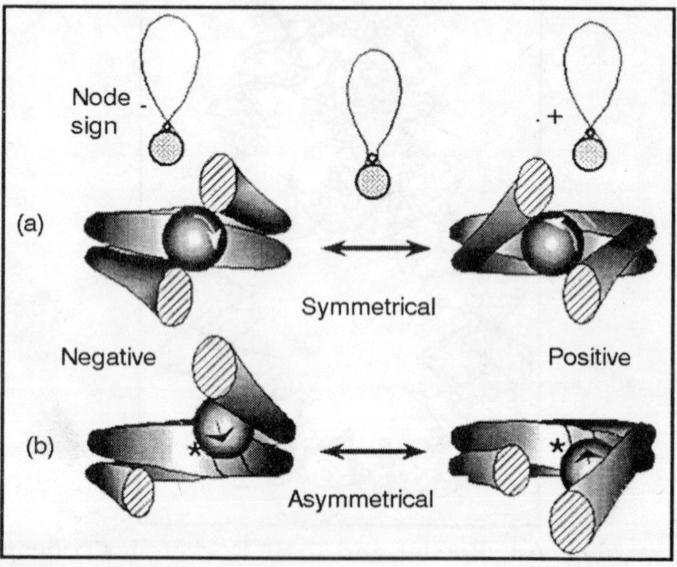

Role of the Ribosome: The route from the DNA code to the protein

Visualizing Cell Biology

Nucleic Acid Nomenclature and Structure

Computational Biology: Bhageerath: an energy based web enabled computer software suite for limiting the search space of tertiary structures of small globular proteins

Molecular modeling of the chromatosome particle

Transcription Regulation in Eukaryotes

Linking structure to function with histone H1

The Endoplasmic Reticulum as a Protein-Folding Compartment: excerpt: Many human diseases, such as cystic fibrosis and Alzheimer's disease, result from improper protein folding. Our studies into the fundamental processes of how the Endoplasmic Reticulum coordinates protein folding and its response to unfolded proteins will provide new avenues to treat the diseases of protein misfolding. Our recent studies also demonstrate, however, that the unfolded protein response signaling pathways are also essential for cell differentiation and cell function under normal physiological conditions.

CHROMOSOMES: GOOD THINGS COME IN VERY SMALL PACKAGES

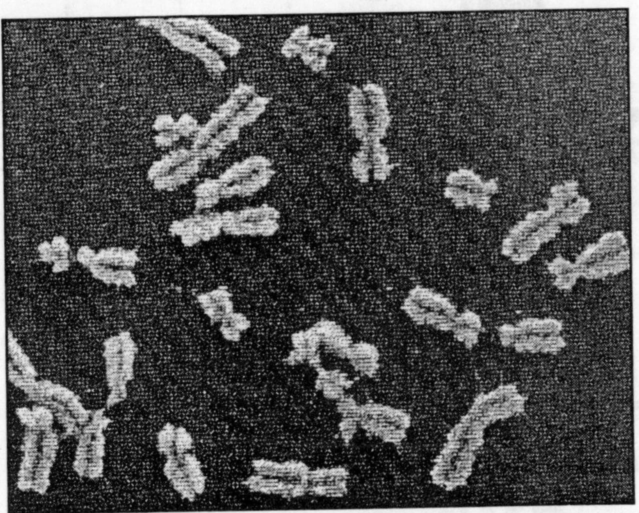

If we unravelled and stretched out the set of DNA molecules that reside in the nucleus of the tiniest human cell, they would extend 5 feet end-to-end. During cell division, however, these molecules tightly wind themselves up into microscopic packages called chromosomes. John Kyrk's website has shown the 3D structure of chromosomes: how the DNA molecule is wound around cylindrical proteins called histones, then further wound into a fibre of chromatin which is then finally wound up to form chromosomes. The information density this packaging achieves is staggering: 1.88 x 10^{21} bits (2 billion gigabytes) of information per cubic centimeter. Oh, and that includes error correction logic. No wonder DNA scaffolding is being looked at as the basis for a possible nanotech storage media.

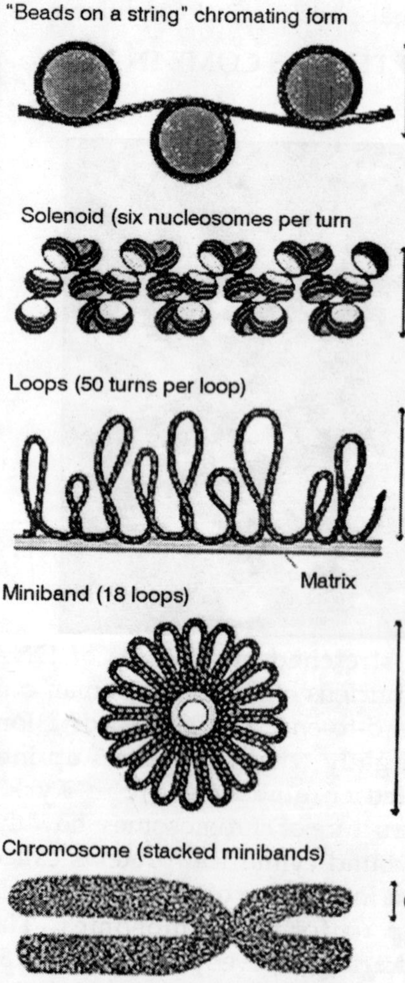

		Base pairs per turn	Packing ratio
DNA double helix	2 μm	10	1
"Beads on a string" chromating form	11 μm	80	6–7
Solenoid (six nucleosomes per turn	30 μm	1200	-40
Loops (50 turns per loop) Matrix	~0.25 μm	60,00C	680
Miniband (18 loops)	0.84 μm	$-1.1\text{x}10^4$	$-1.2\text{x}10^4$
Chromosome (stacked minibands)	0.84 μm	18 loops/ miniband	$1.2\text{x}10^4$

The nucleus of most human cells contains 2 sets of

chromosomes, 1 set given by each parent. Each set has 23 single chromosomes: 22 autosomes [common to both sexes] and an X or Y sex chromosome. (A female will have a pair of X chromosomes; a male will have one X and one Y.)

The picture at the start of this blog entry shows chromosome units that are in the process of duplicating themselves. The symmetric halves of the chromosome are called sister chromatids, both of which contain identical DNA molecules. Each chromatid (and the original chromosome that was split in two to form them) contain two end caps called telomeres and a centromere that forms its waist.

It's amazizng how quickly DNA can wrap itself up into these packages: Fluorescence videomicroscopy and scanning force microscopy were used to follow, in real time, chromatin assembly on individual DNA molecules immersed in cell-free systems competent for physiological chromatin assembly. Within a few seconds, molecules are already compacted into a form exhibiting strong similarities to native chromatin fibers. In these extracts, the compaction rate is more than 100 times faster than expected from standard biochemical assays.

Prior to a cell dividing, the chromosomes are replicated. The sister chromatids that result are initially held together by something called 'chromsome cohesion'. Chromosome cohesion is established during S phase (when the chromosomes are replicated) and is then dissolved completely in metaphase to allow sister chromatids to come apart. The dissolution of cohesion is highly regulated; human cell lines that have defects in the regulation of cohesion show the hallmarks of cancer cells. Furthermore, it has been suggested that the abnormal karyotypes that result in diseases such as Down syndrome are the result of the improper dissolution of chromosome cohesion.

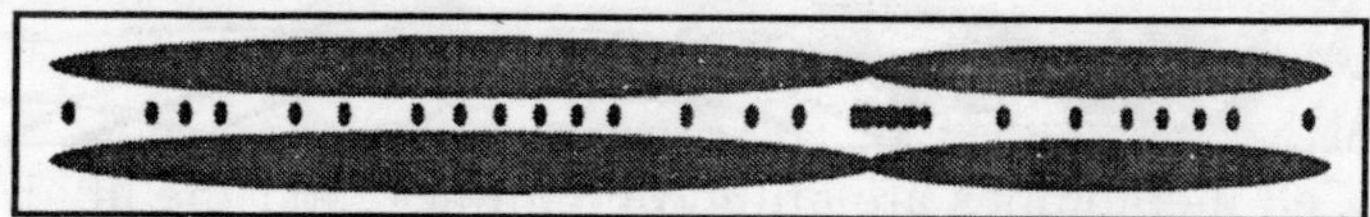

Cohesin sites (red ovals) are concentrated at the centromere/pericentric region, but also occur along the arms of the chromatids.

MIT recently discovered that a protein called MEI-S332 regulates the chromosome cohesion, releasing the chromosomes from each other by adding a phosphate to the binding point.

These findings are particularly significant given that researchers have found that levels of MEI-S332 are higher than normal in 90% of all breast cancers.

According to Clarke, this might mean that when there's too much of the protein, the chromosomes don't separate properly, or it might mean that the MEI-S332 gene is mutated on the chromosomes.

During cell division (mitosis), the two sister chromatids of each of the 46 chromosomes are pulled apart. Protein filaments called microtubules attach the centromeres of the sister chromatids of the chromosomes to the cytoskeleton on opposite sides of the cell. The cytoskeleton is located along the inside of the cell membrane and acts like a nano-scale motor, pulling on the microtubules to line up the chromosomes in the middle of the cell and then literally yanking the sister chromatids apart.

There's also a fascinating movie of a plant cell undergoing mitosis. It took me a while to appreciate that the structural packaging of DNA differs in bacteria, plants and animals.

ION CHANNELS: GATES IN THE CELL WALL

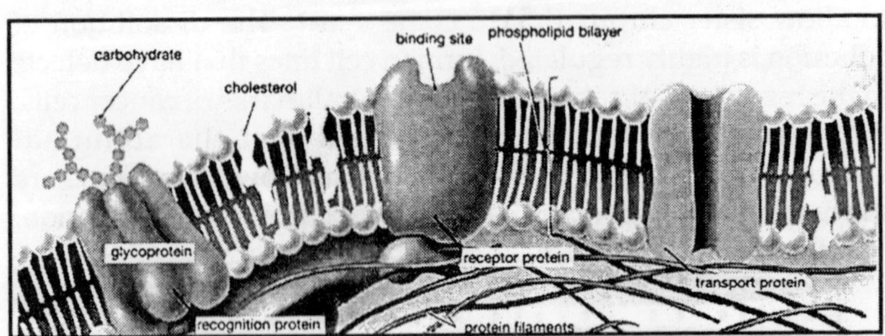

Cell membranes are around 7nm thick, and the lipid bi-layers that they are made of are highly impermeable to ions. The membranes are studded with numerous proteins that have a wide variety of functions. Some of these proteins, for example, form pores that allow molecules to flow from one side of the cell membrane to the other. Hydrophobic amino acids line the outside of the pore, binding the pore into the membrane, and hydrophilic amino acids line the inside of the pore, providing a channel that is friendly to water-soluble chemicals.

The characteristics of the pore are determined by the nature of the protein molecule it is made of. For example, if it is made of a protein called aquaporin, it will create a hydrophilic channel through the cell membrane that has a passage big enough for water molecules to pass through but too small for substances such as alcohol.

Certain types of pore molecules are 'gated' so that they can be shut or opened in response to certain conditions. Pores that allow specific ions to pass through the cell membrane are called 'ion channels'.

"Visualizing a Potassium Channel": Around 50 years ago researchers showed that electrical activity in neurons is produced by subtle changes in the neuron's potassium concentration. "Since then, it's been well established that the flow of potassium ions is central to many different cellular processes," said MacKinnon. Potassium currents in the brain, for example, underlie perception and movement, and the heart's contraction relies upon the steady ebb-and-flow of potassium.

To maintain the correct concentration of potassium, cells are equipped with pore-like proteins that poke through the cell membrane.

These proteins, called ion channels or potassium channels, create sieves through which potassium ions flow from inside to outside the cell.

Channels are just one mechanism to move ions in and out of the cell, however. There are also passive carriers (molecules that enclose ions and move by following chemical or electrical gradients), and active carriers such as ion pumps. The main difference between channels and carriers is that with carrier proteins there is never an open channel all the way through the membrane.

ION CHANNELS

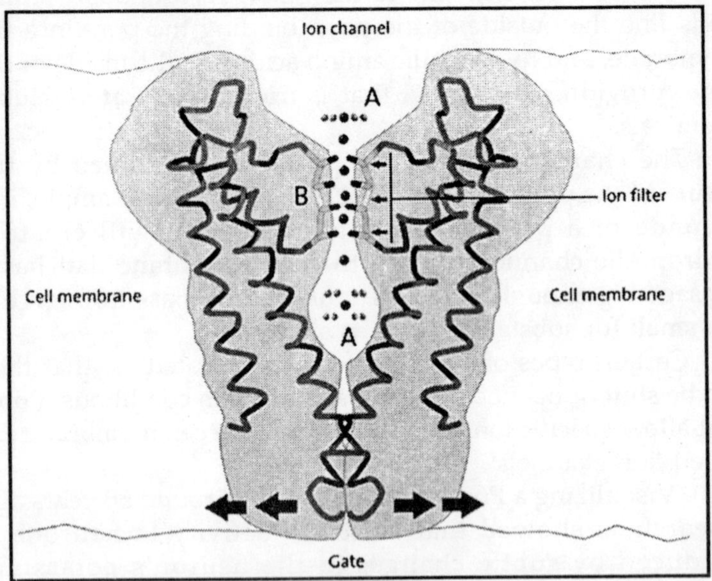

Ion channels have to be able to admit one ion type selectively, but not another, and to open and shut and sometimes to conduct ions in one direction only. But until recently, it was not known how this molecular machinery really worked.

Ion channels: During the 1970s it was shown that the ion channels were able to admit only certain ions because they were equipped with some kind of "ion filter". Of particular interest was the finding of channels that admit potassium ions but not sodium ions – even though the sodium ion is smaller than the potassium ion. It was suspected that oxygen atoms in the transport protein played an important role as "substitutes" for the water molecules with which the potassium ion surrounds itself in a solution of water and from which it must free itself during entry to the channel. Only the Ion can pass, not it hydrated counterpart.

In 1998 Roderick MacKinnon determined the first high-resolution structure of an ion channel and revealed for the first time how an ion channel functions at atomic level... it uses an ion filter, which admits potassium ions and stops sodium ions. His ability with X-ray crystallography made it possible to unravel how the ions passed through the channel. The ions could also be visualized in the crystal structure – surrounded by water molecules just before they enter the ion filter; right in the filter, and when they meet the water on the other side of the filter.

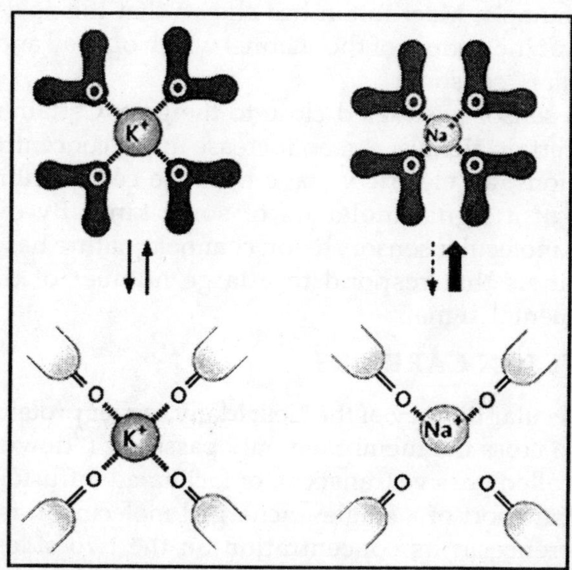

OUTSIDE THE ION FILTER (upper Fig.)

Outside the cell membrane the ions are bound to water molecules with certain distances to the oxygen atoms of the water.

INSIDE THE ION FILTER (lower Fig.)

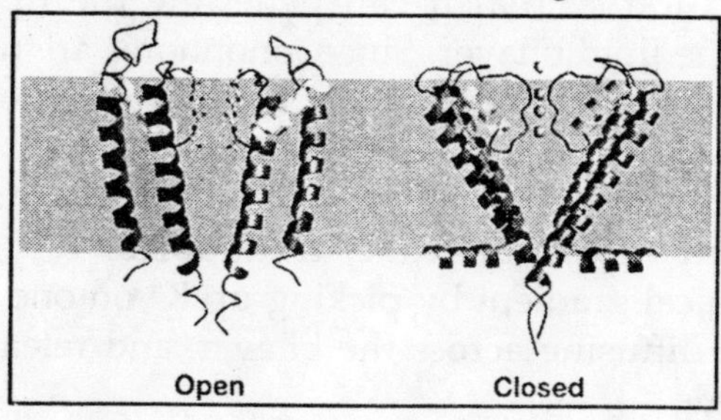

For the potassium ions the distance to the oxygen atoms in the ion filter is the same as in water. The ion channel is able to strip the potassium ion of its water and allow it to pass at no cost in energy. The sodium ions, which are smaller, do not fit in between the oxygen atoms in the filter. This prevents them from entering the channel.

Cells must also be able to control the opening and closing of ion channels. MacKinnon has shown that this is achieved by a gate at the bottom of the channel which opened and closed a molecular "sensor".

This sensor is situated close to the gate. Certain sensors react to certain signals, e.g. an increase in the concentration of calcium ions, an electric voltage over the cell membrane, or binding of a signal molecule of some kind. By evolving different molecular sensors to ion channels, nature has created ion channels that respond to a large number of different environmental signals.

PASSIVE ION CARRIERS

Molecular Biology of the Cell: Many carrier proteins allow solutes to cross the membrane only passively ("downhill"), a process called passive transport, or facilitated diffusion. In the case of transport of a single uncharged molecule, it is simply the difference in its concentration on the two sides of the membrane (its concentration gradient) that drives passive transport and determines its direction.

Ionophores are small hydrophobic molecules that dissolve in lipid bilayers and increase their permeability to specific inorganic ions. They operate by shielding the charge of the transported ion so that it can penetrate the hydrophobic interior of the lipid bilayer. Since ionophores are not coupled to energy sources, they permit the net movement of ions only down their electrochemical gradients.

Valinomycin is an example of a mobile ion carrier. It is a ring-shaped polymer that transports K^+ down its electrochemical gradient by picking up K^+ on one side of the membrane, diffusing across the bilayer, and releasing K^+ on the other side.

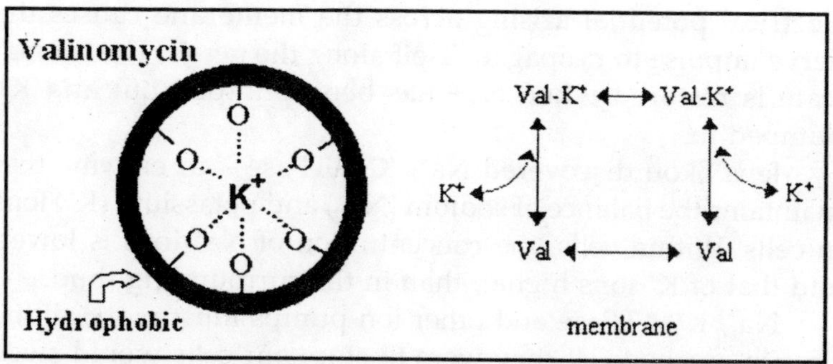

Similarly, FCCP, a mobile ion carrier that makes membranes selectively leaky to H^+, is often used to dissipate the H^+ electrochemical gradient across the mitochondrial inner membrane, thereby blocking mitochondrial ATP production.

A23187 is yet another example of a mobile ion carrier, only it transports divalent cations such as Ca^{2+} and Mg^{2+}. When cells are exposed to A23187, Ca^{2+} enters the cytosol from the extracellular fluid down a steep electrochemical gradient. Accordingly, this ionophore is widely used to increase the concentration of free Ca^{2+} in the cytosol, thereby mimicking certain cell-signaling mechanisms.

ION PUMPS

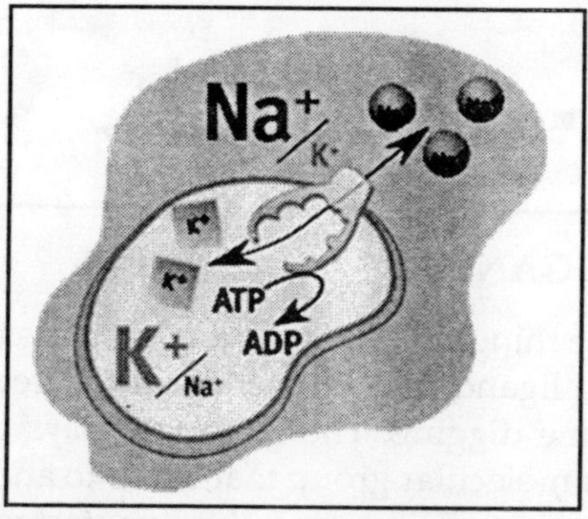

THE FIRST ION PUMP DISCOVERED

Ion composition outside the cell differs from that inside. When e.g. a nerve is stimulated Na^+ flows into the cell. The electrical potential arising across the membrane causes the nerve impulse to propagate itself along the nerve. The normal state is restored when Na+ has been pumped out and K+ pumped in.

Jens Skou discovered Na^+, K^+-ATPase - an enzyme that maintains the balance of sodium (Na^+) and potassium (K^+) ions in cells. Within cells, the concentration of Na^+ ions is lower, and that of K^+ ions higher, than in the surrounding fluid.

Na^+, K^+-ATPase and other ion pumps must work all the time in our body. If they were to stop, our cells would swell up, and might even burst, and we would rapidly lose consciousness. A great deal of energy is needed to drive ion pumps - in humans, about 1/3 of the ATP that the body produces. The Na+, K+-ATPase cycle:

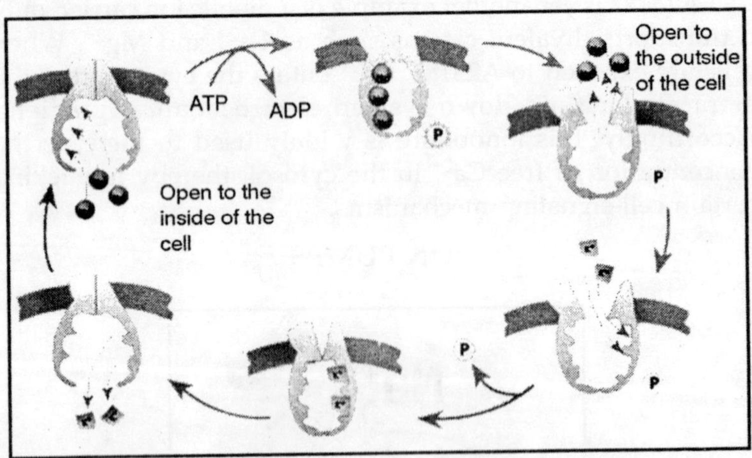

LIFE AND LIGANDS

One of the things we keep coming across is the expression 'ligand', as in 'ligand gated ion channel' or 'receptor ligand'. So we did some digging. The dictionary says it's "An ion, a molecule, or a molecular group that binds to another chemical entity to form a larger complex." But I've started to appreciate that it's a lot more than that. From what we can tell, life happens right at the intersection of two boundaries: the boundary between Newtonian physics and Quantum physics, and the boundary between order and chaos. And the kind of weak, reversible binding between molecules and proteins that we get there is what makes life possible.

Ligands

- The biological function of a protein typically depends on the structure of specific binding sites. These sites are located at the surface of the protein and are determined by geometrical arrangements and physico-chemical properties of tens of non-hydrogen atoms.

The ability of proteins to form specific stable complexes is fundamental to biological existence. The interaction between ligand and protein takes place at the surface of the protein. This surface is very complex and convoluted. Furthermore, bound ligands vary greatly in size and properties. The smallest ligands such as O_2 and NO consist of two covalently linked atoms showing no or only partial atomic charges. Interactions between proteins and ligands of this type are defined by special arrangement of the electron systems of each participant.

Likewise, small ions, e.g. calcium, sodium etc., form a complex compound or similar structure with a few special charged atoms of the protein. A large number of known protein ligands are prosthetic groups, substrates and coenzymes. Their molecular masses lie between 100 and about 2000 Da and their binding sites are larger and more complex. The other end of the size scale is defined by the largest interacting partners of proteins: other biological polymers like nucleic acids, other proteins, and polysaccharides. They show molecular masses from 5000 up to 100,000 Da and more.

The binding sites of smaller ligands seem to be little caves (grooves, pockets, cavities, depressions) at the surface of proteins.

- A ligand refers to a speciûc molecule that can bind to a protein. With respect to viruses, a ligand is a protein on the outer coat of a virus that can bind to a receptor protein on the surface of a cell that the virus will infect. Ligands on the surface of a virus can only bind with speciûc receptor proteins. Different cell types contain different receptor proteins.

- A ligand is simply a molecule which interacts with a protein, by specifically binding to the protein. We should not make the mistake of thinking of a ligand as a relatively small molecule, involved in some obscure biochemical pathway. This narrow view is incorrect. For example consider the case of a repressor protein involved in the regulation of a gene. The repressor protein binds specifically to a section of chromosomal DNA in order to prevent that gene being expressed.

In this case, DNA is the ligand. This example also illustrates another important aspect of protein ligand binding which is that the interaction must be specific. The interior of the cell contains many different potential ligands and the protein must interact only with the appropriate molecule. So a ligand can be a nucleic acid, polysaccharide, lipid or even another protein.

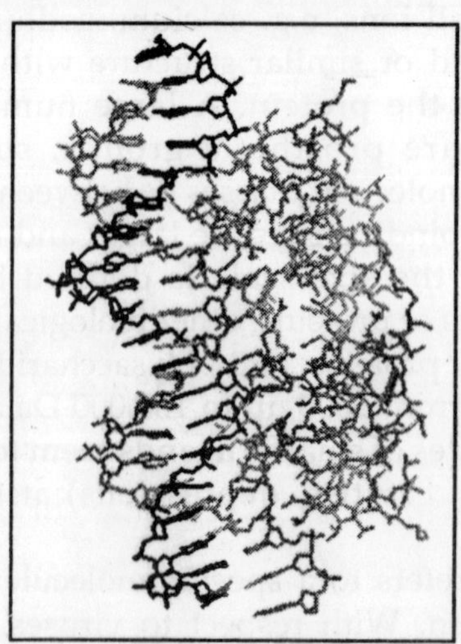

Protein ligand binding is involved in many cell functions including hormone receptors, gene regulation, transport across membranes, the immune response and enzymes catalysis.

Although we may not think of an enzyme catalysed reaction as protein ligand binding, initially the enzyme (protein) must bind to the substrate (ligand) before any chemical changes occur.

An example shown in the diagram at left is a gene regulating protein which binds specifically to DNA preventing transcription. The two chains of DNA on the left of the picture are shaded differently. There are two polypeptide chains shown on the right of the picture. In this case we can see that the protein which binds the DNA ligand is a dimer i.e. composed of two separate polypeptide chains.

Any binding of a ligand to a protein is also reversible. The physical interactions between a protein and ligand are the same as those between the protein and water molecules or hydrogen ions for example but these latter molecules do not bind to the protein at a specific site. The ligand will however bind to the protein at a specific site. This site has the necessary physical characteristics to make ligand binding favourable.

The specificity of a binding site for a particular ligand can vary however depending on the structure of the protein. For example, the protein may be able to bind two different ligands and each of these ligands will then compete for the proteins binding site. Typically, these different ligand molecules will share some structural or physical properties and thus both be able to fit into the binding site on the protein. This is known as competitive binding. In some cases, there may be more than one binding site on the protein molecule. Two different ligand or two similar ligands may be able to bind to the protein, one at each binding site. This allosteric binding is quite common in biological systems. Haemoglobin binding to oxygen is perhaps the most commonly cited example of this. We may already be familiar with allosteric regulation of enzyme action.

SUMMARY OF PROTEIN LIGAND BINDING

- Binding is specific to a particular ligand or group of ligands
- Binding occurs at a particular site in the protein molecule

- Binding is reversible

Electrostatic Steering and Ionic Tethering in Enzyme-ligand Binding

Although the driving force for ligand binding is often ascribed to the hydrophobic effect, electrostatic interactions also influence the binding process of both charged and nonpolar ligands. Electrostatic steering of charged substrates into enzyme active sites conserves electrostatic potential energy localized at the active sites and are the primary determinants of the bimolecular association rates.

A more subtle effect is "ionic tethering: salt links can act as tethers between structural elements of an enzyme that undergo conformational change upon substrate binding, and thereby regulate or modulate substrate binding. Ionic tethering can provide a control mechanism for substrate binding that is sensitive to the electrostatic properties of the enzyme's surroundings even when the substrate is nonpolar

ENZYMES: COME TOGETHER, RIGHT NOW, OVER ME

Definition of an *enzyme*: Any of numerous proteins or conjugated proteins produced by living organisms and functioning as biochemical catalysts. Enzyme names almost all end in "-ase".

Definition of *Catalyst*: a substance that decreases the required activation energy to increase the reaction rate. Chemical composition of a catalyst is not altered by the reaction and thus a single catalyst molecule can be used over and over again. Definition of *Kinase*: Any of various enzymes that catalyze the transfer of a phosphate group from a donor, such as ADP or ATP, to an acceptor. Kinases alter the shape of acceptor proteins and this change in shape changes which binding sites on the protein molecule are exposed and which other proteins are able to bind with the available sites.

Enzymes are life's Catalysts

Tears convey happiness, or sadness, but they do

something else as well, and exactly what was accidentally discovered in the early 1900s. A bacteriologist by the name of Alexander Fleming happened to create an inadvertent experiment when a teardrop fell into one of his bacteria cultures. It set the stage for our modern understanding of how cell processes are controlled.

Many years later the chemical in tears was isolated and described. It was an enzyme — now called Fleming's Lysozyme. Lyse means to break apart, zyme means enzyme.

We now know how Flemming's Lysozyme works. First, like all enzymes, it's a protein, a huge molecule built from a perfectly ordered assemblage of amino acid building blocks. Its chemical structure fits molecules in the bacterial cell wall like a lock and key, and once the enzyme locks on, the bacterium's cell wall comes apart.

Enzymes often aren't assembled by our cells until they are needed for a specific job. How does a cell know when turn on a gene for making a particular enzyme?

We know something about this turning-on of genes from experiments with bacteria. The selected subject for study: yogurt. Yogurt is really milk, processed by some friendly bacteria. Breaking down the milk is accomplished by specific enzymes the bacteria produce.

Bacteria break down milk sugar, lactose, using an enzyme they produced called galactosidase.

When there is no lactose around the bacteria don't bother making galactosidase, but if they encounter molecules of lactose, they start producing the enzyme.

On the bacterium's DNA is the gene for the galactosidase enzyme. The gene has a start signal. Hanging on to it is a protein called a repressor. The repressor's molecular structure recognizes the lactose molecule.

When a lactose molecule bonds on, the repressor releases its grip. RNA polymerase can then slip in and transcribe the galactosidase gene. Soon the bacterial ribosomes are turning out enzymes that will break down the lactose into something the bacterium can use. In the process the bacteria create a tasty treat for us.

Biology online's Enzymes and Chemical Reactions

Metabolism consists of synthesis (anabolism) and breakdown (catabolism) of organic molecules required for cell structure and function.

Chemical reactions involve:

- The breaking of chemical bonds in reactant molecules
- Making of new chemical bonds to form product molecules. Energy is either added or released as heat during chemical reactions.

Enzymes are *protein catalysts.* (A few RNA molecules also possess catalytic activity). In an enzyme-mediated reaction, an enzyme binds to reactants (substrates) to form an enzyme-substrate complex, which breaks down to release products and the enzyme. The region of the enzyme to which the substrate binds is called the active site, the shape of which determines the chemical specificity of the enzyme.

Reversibility of a Reaction: Energy released during a reaction determines reversibility of a reaction. Greater the energy released during a reaction, smaller is the probability of product molecules obtaining this energy and undergoing the reverse reaction to reform the reactants. In such a case, ratio of product to reactant concentration will be large and the reaction will tend to be irreversible. In a reversible chemical reaction, rate of forward reaction decreases and rate of reverse reaction increases as the reaction progresses until the two are equal in a state called the chemical equilibrium, at which point there is no further change in the concentration of reactants and products.

Cofactors

Substances that bind to enzymes to alter their conformations and make them active. Some cofactors are trace elements. In cases where the cofactor is an organic molecule, it is called a coenzyme. Coenzymes are derived from vitamins.

Vitamins

From the National Science Teachers: Vitamins are organic molecules that function in a wide variety of capacities within the body. The most prominent function is as cofactors for enzymatic reactions. The distinguishing feature of the vitamins is that they generally cannot be synthesized by mammalian cells and, therefore, must be supplied in the diet.

ATP: Power to the people, right on!

$$
\begin{array}{c}
\text{ATP: } O^- - P(=O)(O^-) - O - P(=O)(O^-) - O - P(=O)(O^-) - O - CH_2 - \text{ribose (OH, OH)} - \text{adenine } (NH_2)
\end{array}
$$

Fig. The ATP Molecule

ATP is regarded as a universal source of energy in all kinds of cells. It is produced mainly in the oxidizing of energy-rich (reduced) compounds in the course of the respiratory chain and in photosynthesis.

ENZYMES ARE LIFE'S CATALYSTS

Adenosine triphosphate is the primary molecule to which energy is transferred during the breakdown of fuel molecules, carbohydrates and fats. (A cell cannot use heat energy to perform its functions). ATP is then hydrolyzed to release energy which can be used by the energy requiring processes in cells such as production of force and movement, active transport across membranes and synthesis of organic molecules used in cell structures and functions.

ATP is an energy transfer molecule and NOT an energy storage molecule. It transfers energy from fuel molecules to cells in small amounts

- When energy (such as in a photon) is pumped into a chemical system, the energy partitions into thermal and electronic components. The thermal component makes the molecules move faster, and the electronic component increases the number of "high-energy" electronic states. Both energy components will foster molecular organization: the faster the molecules vibrate, rotate, and translate, and the more of them that are in electronic states above ground level, the higher is the probability that the molecules will interact and the more work can be done in organizing them.

Living organisms store photon energy in chemical form, and then trickle it down molecular chains to the individual molecular bonding sites.

The photon energy is stored in the covalent bonds of glucose – about 6 quanta of photon in one glucose molecules. From this reservoir, energy then flows along various pathways, nursing everything, all organization and all work. The chemical energy chains that nurse macromolecular organization commonly use ATP as their final link.

- Trudy Wassenaar: Many different proteins extract energy from the hydrolysis of ATP. All of them appear to couple some sort of motion of the protein with the energy-releasing hydrolysis. The motion of the protein can in turn be harnessed to do necessary things, such as pump chemicals across a cell membrane, move the cell around, synthesize new proteins, and so on.
- Remember the Ion Pump description that won Jens C. Skou the 1997 Nobel prize? Well, he shared the 1997 Nobel Prize in Chemistry with Paul D. Boyer, John E. Walker "for their elucidation of the enzymatic mechanism underlying the synthesis of adenosine triphosphate (ATP). Check out Boyer's lecture "Energy, Life and ATP", and Walker's lecture"ATP Synthesis by Rotary Catalysis". Digging further back in time brings we to the 1953 The Nobel Prize in

Physiology or Medicine awarded to Hans Krebs for his work on the Citric Acid Cycle, and Fritz Lipmann for his discovery of "co-enzyme A and its importance for intermediary metabolism". Lipmann talks about "energy-rich phosphate bonds" and "transformations of electron transfer potential to phosphate bond energy".

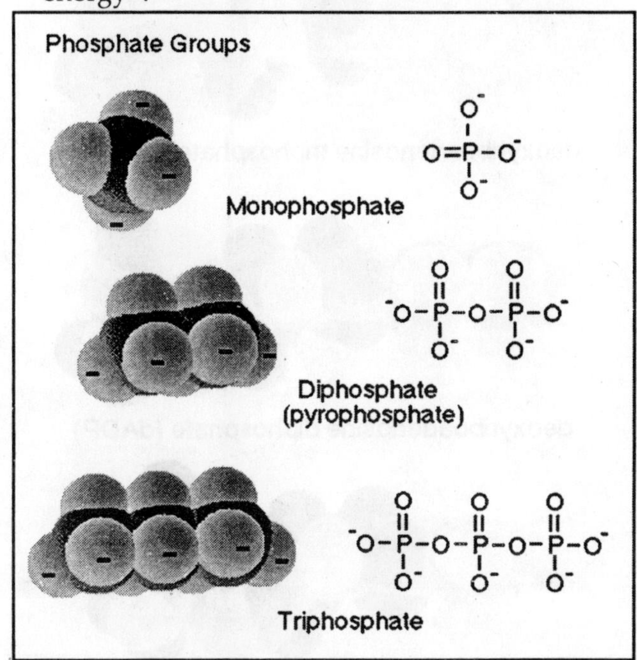

- University of Connecticut: When phosphate groups are joined together, they have a strong tendency to repel each other, because of the high concentration of negative charge in the very polar and usually ionized oxygen atoms. [the O^- in the ATP molecular diagram] As a result, molecules with two or three phosphate groups are good energy donors, readily releasing energy along with the transfer of phosphate groups.

It feels like a number of things that have been discussed in these notes are finally starting to come together. Where else have we seen these phosphate groups before? Oh yeah - the nucleotides that make up DNA.

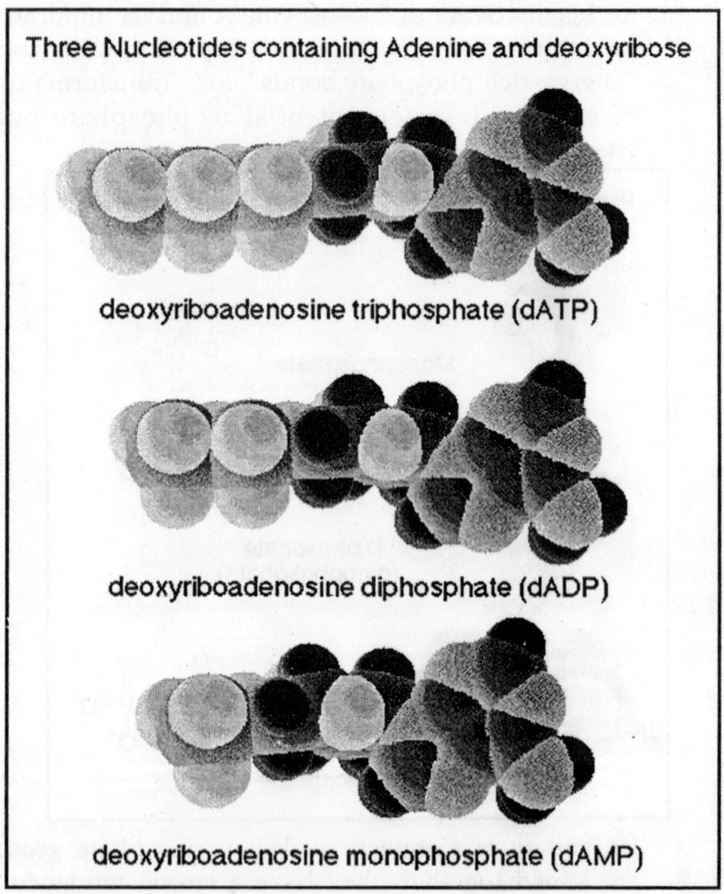

It's not hard to imagine the ionized O^- atoms associated with all of those phosphate groups kicking the molecules around energetically, making sure that they encounter the other molecules that they need to form bonds and ligands with. And why don't these O^- atoms bond with, say Na^+ or K^+ ions? For the same reason that we don't find too many NaCl crystals in the body - they are dissolved by water.

Besides the adenosine nucleotide phosphates occur also uracil, cytosine and guanine phosphates:

UMP, UDP, UTP, CMP, CDP, CTP, GMP, GDP, GTP.

The triphosphate nucleosides of the mentioned compounds including ATP are components of RNA. They are integrated into the polymer by cleavage of pyrophosphate (= PP).

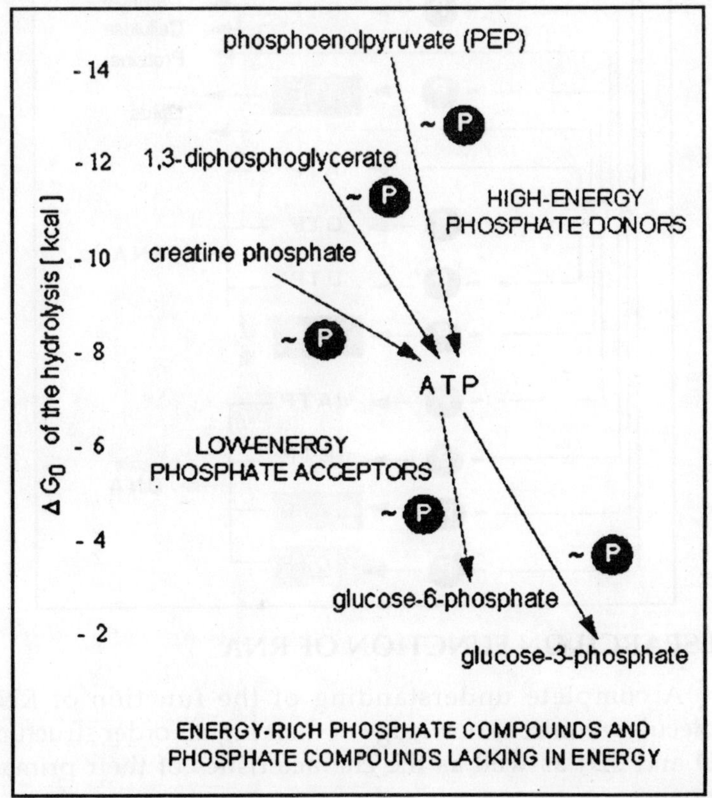

The corresponding desoxyribose derivatives (dATP, dGTP, dCTP....) are necessary for DNA synthesis where dTTP is used instead of dUTP. The terminal phosphate residues of all nucleoside di- and triphosphates are equally rich in energy. The energy set free by their hydrolysis is used for biosyntheses. They share the work equally: UTP is needed for the synthesis of polysaccharides, CTP for that of lipids and GTP for the synthesis of proteins and other molecules.

These specificities are the results of the different

selectivities of the enzymes that control each of these metabolic pathways.

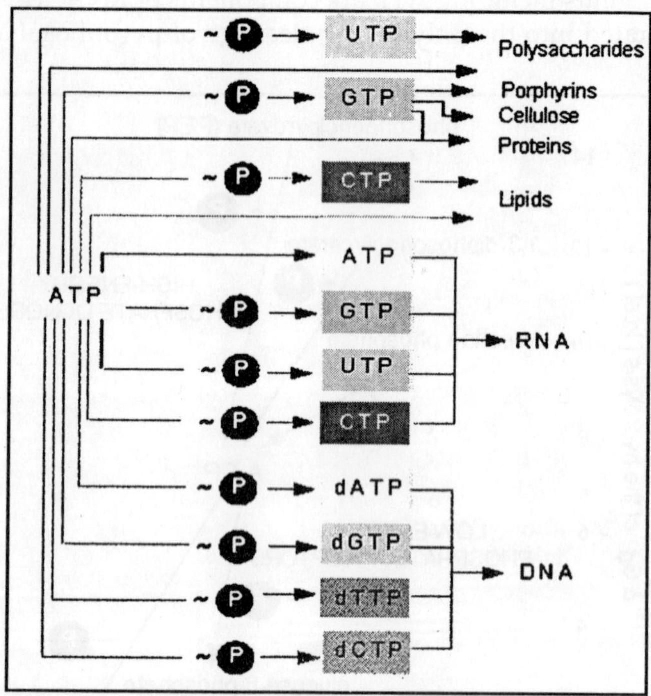

RESEARCH ON FUNCTION OF RNA

A complete understanding of the function of RNA molecules requires knowledge of their higher order structures (2D and 3D) as well as the characteristics of their primary sequence.

RNA structure is important for many functions, including regulation of transcription and translation, catalysis, transport of proteins across membranes and the regulation of RNA viruses.

The understandings of these functions are important for basic biology as well as for the development of drugs that can intervene in cases where pathological functionality of these molecules occurs.

Our group does research and development of

methodologies for improving RNA folding and analysis techniques to help further our understanding of the functional properties of these molecules.

In addition, we are focusing on the emerging field of RNA nanobiology. RNA represents a relatively new molecular material for the development of biologically oriented nano devices. It is an interesting material because of its natural functionalities, its ability to fold into complex structures and self assemble. We have developed computational methodologies that permit the design of RNA based nano-particles that potentially have a variety of uses. Thus, our research on RNA covers four highly related and integrated areas of computational research;

- Research in algorithms for RNA secondary structure prediction and analysis;
- RNA biology and its relationship to sequence and secondary structure folding characteristics; 3) Research in algorithms for RNA
- 3D structure prediction and analysis and their application to RNA biology;
- Research in algorithms for the design and analysis of RNA nanoparticles. What is learned in one area is applied to the other areas, enhancing our understanding of RNA structure, function, and RNA nanobiology and self-assembly.

Parallel/Heterogeneous Computational Biology and RNA Structure

Revolutionary changes in computational paradigms are required to maintain the necessary computational power to solve problems in molecular biology. Methodologies based on sequential computer architectures cannot be expected to continually keep pace with the needed computational speeds. In order to accommodate the high speeds that are necessary, heterogeneous and highly parallel computational techniques are required. Our group was one of the pioneers in the area of computational biology and the use of parallel high performance computer architectures for this endeavor.

Parallel/Heterogeneous Computation and RNA Structure

We were the first to develop an RNA folding technique that uses concepts from genetic algorithms. Our algorithm, MPGAfold, was originally developed to run on a massively parallel SIMD supercomputer, a MasPar MP-2 with 16384 processors. This algorithm was modified and now runs on parallel MIMD high performance computers which include multicore Linux clusters. Exceptional scaling characteristics are obtained with the ability to run the algorithm with hundreds of thousands of population elements. RNA pseudoknot prediction is part of the GA, resulting in its ability to predict tertiary interactions.

Other features include the ability to incorporate different energy rules, and the forced inhibition and embedding of desired helical stems. In addition, STRUCTURELAB, our heterogeneous bioinformatical RNA analysis workbench can be used in conjunction with MPGAfold and RNA2D3D to produce predicted 3D atomic coordinates of RNA structures along with the visualization of these structures. Also, we developed a novel interactive visualization methodology that is part of STRUCTURELAB. This technique enables the comparison and analysis of multiple sequence RNA folds from a phylogenetic point of view, thus allowing improvement of predicted structural results across a family of sequences.

We developed one of the best algorithms, KNetFold, for RNA structure prediction from sequences alignments. The algorithm uses a unique hierarchical classification network based on mutual information, thermodynamics and Watson-Crick base-pairedness to predict structures. In addition, we have developed a web based application, CorreLogo, that uses mutual information derived from RNA sequence alignments to determine covariations amongst base-paired positions. The algorithm includes a unique error measure and depicts results in 3D.

Computational Studies of RNA Folding Pathways

RNA folding pathways are proving to be quite important

in the determination of RNA function. Studies indicate that RNA may enter intermediate conformational states that are key to its functionality. These states may have a significant impact on gene expression. It is known that the biologically functional states of RNA molecules may not correspond to their minimum energy state, that kinetic barriers may exist that trap the molecule in a local minimum, that folding often occurs during transcription, and cases exist in which a molecule will transition between one or more functional conformations before reaching its native state.

Thus, methods for simulating the folding pathway of an RNA molecule and locating significant intermediate states are important for the prediction of RNA structure and its associated function. Several biological RNA folding pathways have been successfully studied using MPGAfold and STRUCTURELAB. Examples include potato spindle tuber viroid, the host-killing mechanism of Escherichia coli plasmid R1, the hepatitis delta virus and HIV. Computational results are consistent with those derived from biological experiments and novel structural interactions and important functional intermediate and native states have been predicted. These have lead to further successful confirmatory experiments.

Computational Studies of Three-Dimensional RNA Structures

Some structural elements of RNA molecules have been studied using molecular mechanics and molecular dynamics simulations. The structures examined included an RNA tetraloop where temperature-dependent denaturation of the tetraloop and the subsequent refolding to the original crystal structure were performed. A three-way junction from the core central domain of the 30S ribosomal subunit from Thermus thermophilus was explored. It has been experimentally determined that the intermolecular interactions between the three-way junction and the S15 ribosomal protein initiate the process of the assembly of the 30S ribosomal subunit. By using molecular dynamics simulations we obtained insights into the conformational transitions of the junction associated with the binding of S15. We have also examined the pseudoknot domain of telomerase. Molecular modeling and molecular dynamics of the pseudoknot domain including its hairpin loop were performed.

Results indicated how the hairpin loop dynamics affected the opening and closing of the non-canonical U-U base pairs found in the stem. The opening suggested nucleation points for the formation of the pseudoknot.

We have also examined the effect of dyskeratosis congenita (DKC)mutations in the loop and how they reduced the propensity for the opening of the stem by forming a relatively stable hydrogen bond network in the hairpin loop. We modeled the pseudoknot itself using our RNA2D3D software combined with phylogenetic analysis. We studied the dynamical impact of the DKC mutations on the pseudoknot with the result that the pseudoknot became unstable while the hairpin form became more stable.

We also recently discovered and elucidated the 3D structure of a new type of translational enhancer that is found in the 3' UTR of the Turnip Crinkle Virus (the first of its kind found). This was accomplished with the combined use of MPGAfold and our 3D molecular modeling software RNA2D3D.

In addition, we have employed methods based on elastic network interpolation to reduce the computational costs related to RNA 3D dynamics. Three-dimensional dynamics trajectories can be determined using a reduced atom representation and given conformational states. Compute time can be reduced from weeks to hours using this approach.

RNA Nanobiology

As previously indicated, RNA nanobiology represents a new modality for the development of nanodevices which have the potential for use in a number of areas including therapeutics. We developed several computational tools that provide a means to determine a set of nucleotide sequences that can assemble into a desired nano complex. One of these tools is a newly developed relational database called

RNAJunction. The database contains structural and sequence information for all known RNA helical junctions and kissing loop interactions.

These motifs can be searched for in a variety of ways, providing a source for RNA nano building blocks. Another computational tool, NanoTiler, permits a user to interactively construct specified RNA-based nanoscale shapes. NanoTiler provides a 3D graphical view of the objects being designed and provides the means to work interactively on the design process even though the precise RNA sequences may not yet be specified, and an all-atom model is not available. It is easy to change the scale on which one works. NanoTiler can use the 3D motifs found in the RNAJunction database with those derived from specified RNA secondary structure patterns to build a defined RNA nano shape. Or, a combinatorial search can be applied to enumerate structures that would not normally be considered.

KINETICS AND MECHANISMS OF VIRAL MEMBRANE FUSION

- Membrane protein oligomerization and mobility
- Inhibitors and antibodies as conformational probes
- Binding and fusion

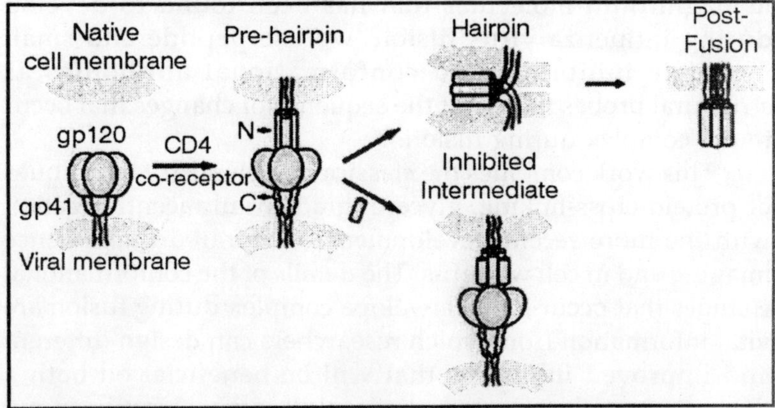

The enveloped viruses including HIV gain access to human cells using an envelope protein complex located on the surface of the virion. In the case of HIV, the transmembrane portion of this protein complex, gp41, is the fusion protein. The globular envelope subunit, gp120, is non-covalently associated to the transmembrane portion and makes initial contact with the cellular receptors.

There is evidence that the envelope protein subunits form a trimer of heterodimers that resembles a stalk structure on the virus surface and on the cell surface of infected cells. The structure of the portion of gp41 that protrudes from the viral surface has been solved in vitro. It is a trimer that forms a very stable six-helix bundle.

This striking stability has led researchers to posit that the structures solved in vitro are representative of the post-fusion form of gp41 with the energy from the conformational change driving fusion. There are numerous peptide inhibitors and conformational antibodies that bind to gp41 only at different stages during the fusion process. This also suggests that there are distinctly different conformations of gp41 that occur as fusion proceeds. It is my hypothesis that gp41, during formation of a fusion pore, undergoes conformational changes that are dramatic and involve both local structural changes and also changes in oligomerization state. If this is the case, it would be similar to the aggregate of at least eight hemagluttinin molecules that has been found to develop during influenza virus fusion. we use peptide and small molecule inhibitors and conformational antibodies as structural probes to dissect the sequence of changes that occur to the complex during fusion.

This work combines the classical biochemistry techniques of protein cross-linking, glycerol gradient ultracentrifugation with the more recent developments of infrared fluorescence imaging and in cell westerns. The details of the conformational changes that occur in the envelope complex during fusion are vital information from which researchers can design different and improved inhibitors that will be beneficial on both a therapeutic and a prophylactic level. It is my objective in my career to provide vital basic research in virology and cell biology in order to merge in vitro structural studies with increasingly greater resolution of protein machines in their native cellular environment.

- Membrane fusion protein reconstitution
- Targeting cancer cells
- Targeting viruses

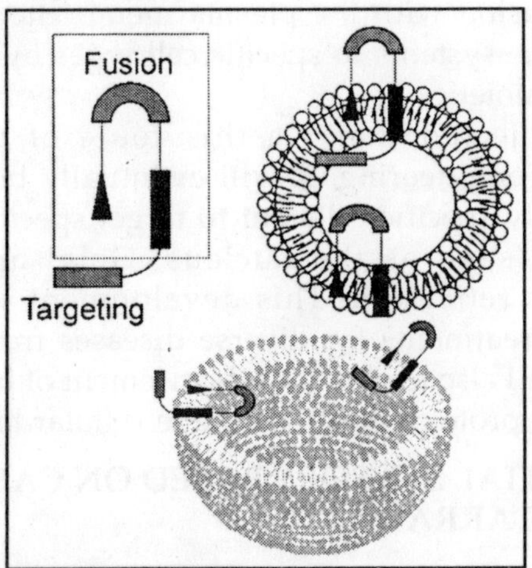

Liposomes have multifunctional capacity which makes them very promising tools in the treatment of cancers, HIV/AIDS, and in DNA delivery for genetic disorders. One of the advantages of liposomes is that they can compartmentalize hydrophobic components in their lipid bilayers along with being able to encapsulate hydrophilic components. This gives researchers the possibility to incorporate different modalities into one formulation including molecules for targeting, biomarker detection, in vivo imaging and delivery of chemotherapy agents.

One of the major drawbacks in using liposomes as drug carriers, however, is that the entry mechanism and the subsequent fate of liposomes after entry into cells is not known from the outset. Many liposomal formulations act simply as transfection reagents by binding molecules of positive charge, such as DNA, to their phospholipid head groups. The DNA is the transferred by an indeterminate merging of the lipid and the DNA molecules. Many of the other types of liposomal formulations are taken up by endocytosis and shuttled unpredictably to intracellular vesicles which might have different chemical compositions and therefore lead to unpredictable delivery. we are overcoming this drawback by developing innovative nanoparticles that deliver therapeutic agents directly into the cytosol of targeted cells utilizing direct membrane fusion with the plasma membrane. we am also targeting these systems to specific cell types by the addition of targeting moieties.

It is projected that via this route of protein and nanoparticle engineering, it will eventually be possible to target not only specific cells but to target specific organelles within cells such as the nucleus, Golgi apparatus or endoplasmic reticulum. This development will provide prospective treatments for diverse diseases from viruses to cancer and will also lead to the advancement of basic research in membrane proteins and membrane cellular biology.

BIOPOTENTIAL SENSORS BASED ON CARBON NANOTUBE ARRAYS

The project goal is the development of a sensor based on carbon nanotubes (CNT) to measure electrical biosignals such as electroencephalography (EEG). The advantage of such an approach is to avoid the use of traditional electroconductive gels. These gels are used to decrease the contact impedance between the sensor and the skin to enhance the signal to noise ratio of the biosignal. The use of these gels often needs skin preparation and the whole process can take up to several minutes per electrode.

It does also cause difficulties in long term monitoring applications because the degradation of the conducting gel properties with time due aging. Furthermore, the necessity of these gels hinders the development of biopotential signal monitoring applications outside of the medical field. Avoiding the use of these gels can open the door a plethora of applications to biosignal monitoring, i.e. EEG, for fatigue or stress. This technology will be very interesting to the transport sector by developing systems that warn operators when drowsiness sets in.

CNT have attracted a huge interest in the research community due to their unique properties: good conductivity, excellent mechanical properties and chemical stability. CNT attached to substrates will provide 'nanobrushes', which will be used to directly contact the skin.

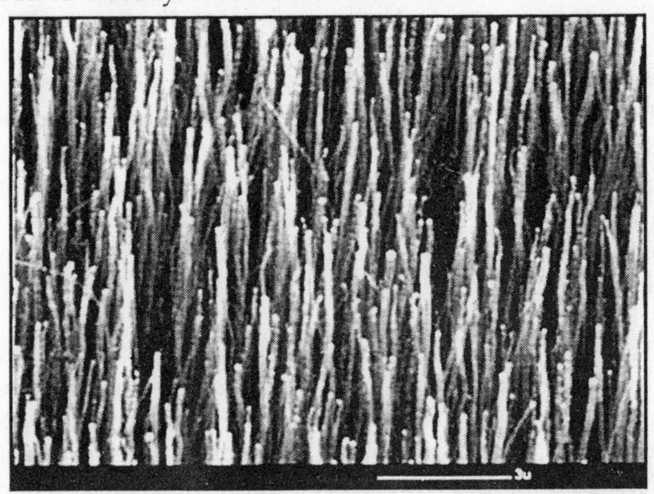

Fig. Aligned Carbon Nanotube array grown by Plasma Enhanced Chemical Vapour deposition

The nanotubes touching the skin surface will accommodate the topography of the skin giving unrivalled surface contact. Furthermore, due to the size of the nanotubes being much smaller than the pores of the skin, they will penetrate the outer layer of dead cells of the skin, the Stratum Corneum (SC). This layer is the origin of the high impedance of the skin. Thus, the CNT brushes will contact the inner epidermis layers giving a low contact resistance ending with the necessity of the electroconductive gels.

BIOSENSING WITH CARBON NANOTUBES

Biomolecules, like DNA and proteins, represent natural nanoscale engineering. They have structure at the sub-nanometer level and have evolved over billions of years to perform dynamic tasks (target identification, target transformation, transport, and electrical conduction) in the living cell. The overall aim of this project is to directly couple proteins to nanoparticles (namely carbon nanotubes) to construct bioelectronic devices. Such devices may be used in biosensor applications, lab-on-a-chip devices or even as drug-delivery systems.

Biosensors are analytical devices that use specific biochemical reactions mediated by isolated enzymes, antibodies/antigens (immunosystems) tissues, organelles or whole cells to detect (and quantify) chemical compounds, usually by electrical, thermal or optical signals Biosensors combine the selectivity of biology with the processing power of modern microelectronics and optoelectronics to offer powerful new analytical tools with major applications in medicine, environmental diagnostics and the food and processing industries.

The idea of the project is to use carbon nanotubes for creating biosensors. These can be functionalised to interact with the different biomoiecules to sense and give a specific response to a certain molecule. There are two ways for modifying nanotubes with target organic molecules: covalently and non-covalently. For the covalent linkage, we use the carboxylic acid functions created by the damaging process of purification of the carbon nanotubes. The non-covalent functionalization is supposed to work thanks to stacking/ hydrophobic interactions. p-p

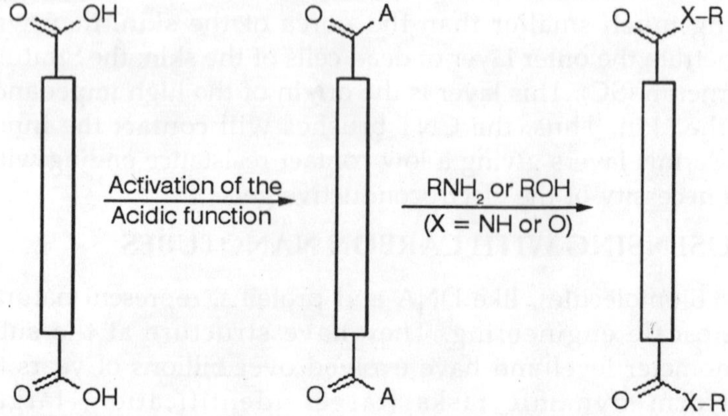

Fig. Diagram of the Covalent Modification Route.

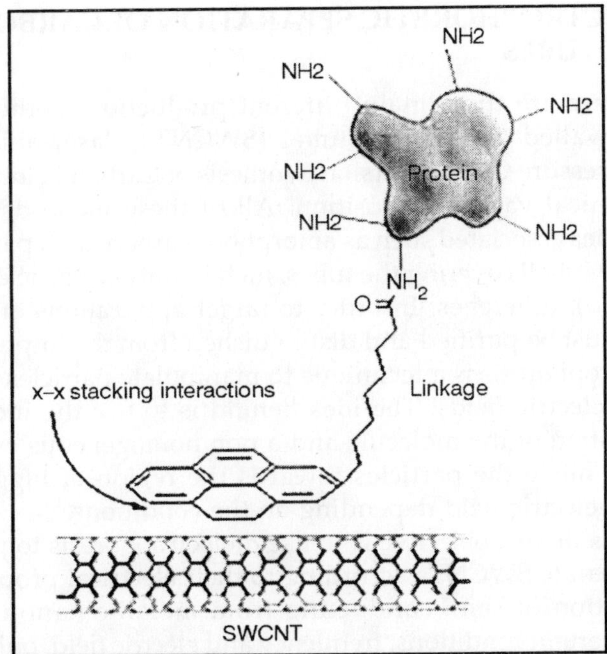

Fig. Diagram of the Non-covalent Modification Route.

Current work involves attaching fluorescence tagged proteins to carbon nanotubes through the use of non-covalently associated linker molecules. The carbon nanotubes are grown into mats and the electrical properties of the modified mats are currently under investigation.

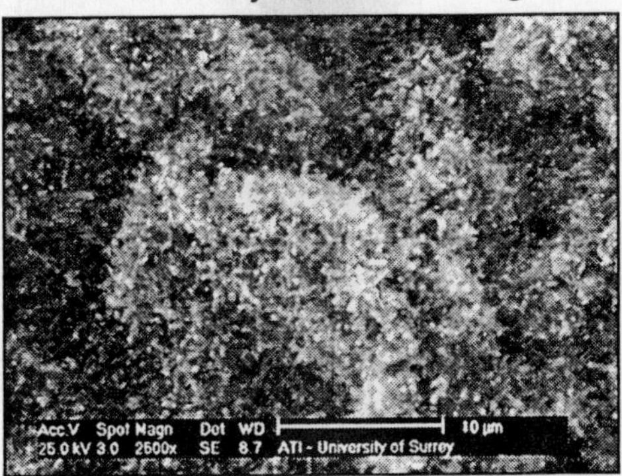

Fig. Scanning electron micrograph of a carbon nanotube mat.

DIELECTROPHORETIC SEPARATION OF CARBON NANOTUBES

There are three main different production methods of single walled carbon nanotubes (SWCNTs): laser ablation, High-Pressure CO Conversion Synthesis of Carbon Nanotubes or chemical vapour deposition. All of these methods have impurities associated such as amorphous carbon (a-C) particles or an a-C shell covering the tubes, metal catalytic particles (Ni, Fe or Co), fullerenes. In order to target applications of CNT these must be purified and distinguished from the impurities. Dielectrophoresis is a technique to manipulate particles using an AC electric fields. The idea behind is to use the induced polarisation of the molecule and a non homogeneous electric field to move the particles towards the region of bigger or smaller electric field depending on the conditions.

This project objective is to use dielectrophoresis to purify and separate SWCNTs according to their electrical properties (separation of semi-conducting from metallic nanotubes). Under certain conditions, frequency and electric field, only one type of tubes are attracted towards the electrodes, leaving the other in solution.

Figure shows swcnts in triton solution, trapped between two gold electrodes and aligned along the electric field.

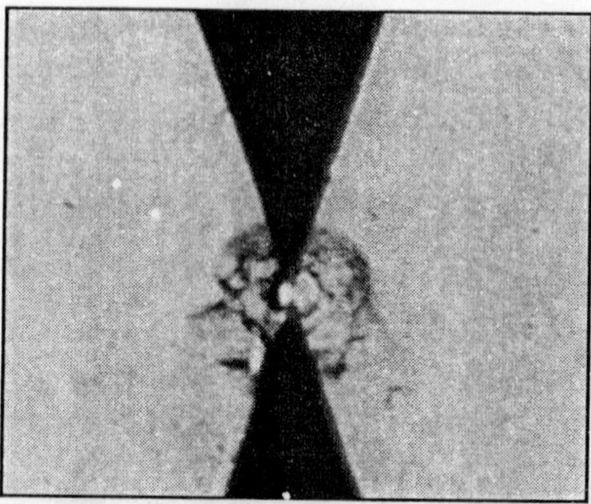

Fig. Optical Micrograph of Carbon Nanotubes Gathered in between two Metal Electrodes using Dielectrophoresis.

Figure shows an Atomic Force Microscopy (AFM) image of a SWCNT rope formed in between the electrodes during the dielectrophoretic process. The samples from macroscopically elongated structures due to the local electrical field enhancement during the process.

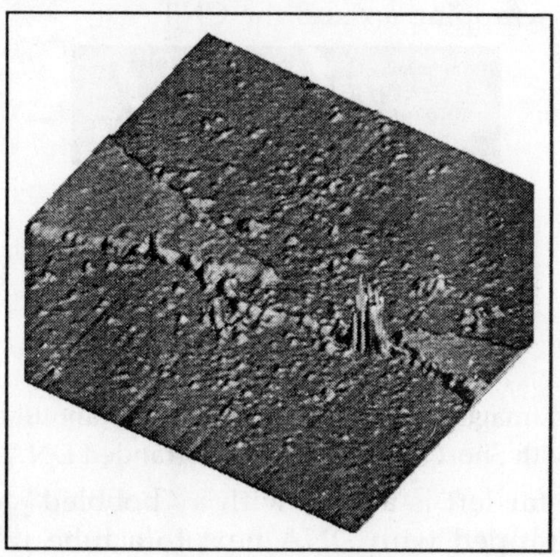

Fig. AFM Image Showing a SWCNT Rope Formed during the Dielectrophoresis Process.

ELECTRONICALLY DETECTING DNA HYBRIDISATION FOR USE IN NOVEL BIOSENSORS

Convenient, cheap, reliable and accurate biosensors are much sort after in molecular biology to detect biological molecules. However, traditionally there has been an intractable problem of transducing the binding event of biomolecule to a detectable signal. Nanotechnology gives us new tools with which to probe the nanometer sized world in which biomolecules react. In particular, carbon nanotubes (CNT) have excellent prospects for being able to probe the activities of biomolecules.

CNT are nanometre sized long, thin cylinders of carbon with extraordinary electronic, thermal and structural properties. They are excellent conductors of both electricity and heat, and have superb tensile strength. But despite their promise, carbon nanotubes remain difficult to work with in a biological context as they are insoluble in water and have no functional groups. To overcome this, the tubes can be made soluble by wrapping them in short strands of single stranded DNA. Figure shows an atomic force microscope (AFM) image of the DNA wrapping around the CNT.

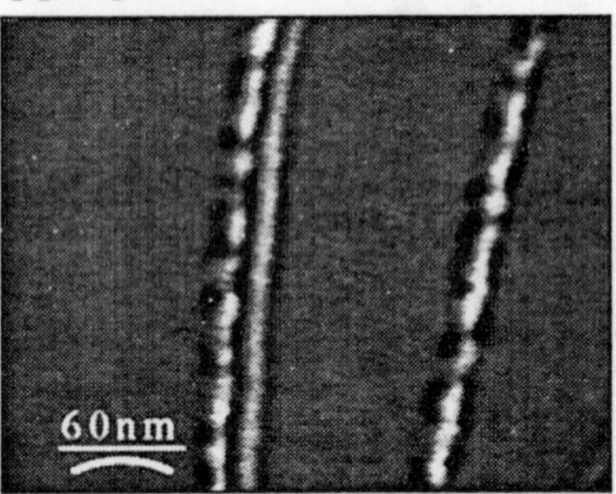

Fig. An AFM Image of Single Walled Carbon Nanotubes Wrapped with Short Strands of Single Stranded DNA.

On the far left is a tube with a "bobbled" appearance which is wrapped with DNA next to a tube which is not wrapped. Using DNA not only solubilises the tubes in water but also has much potential for functionalising the surface of the tubes for use as DNA hybridisation biosensors. We have also been investigating the use of dielectophoresis to position and trap DNA between electrodes with an aim of electronically measuring DNA hybridisation events.

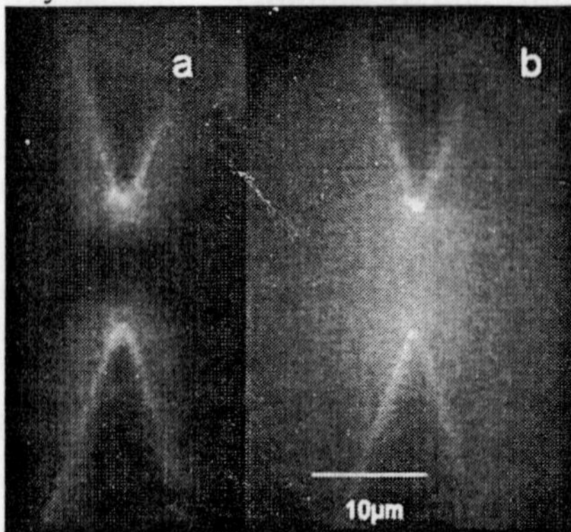

Fig. Dielectrophoresis of Fluorescently Labeled DNA.

Figure shows DNA fluorescently labeled being pulled between two electrodes. This technique uses an electric field to induce a dipole on the DNA. (a) shows the moment when the field is turned on and (b) shows the DNA fully stretched between the electrodes. This has the potential to position DNA in solution to a complementary strand where a hybridisation event could occur and be detected electrically.

GAS ADSORPTION PROPERTIES OF CARBON NANOTUBE ROPES

The conduction properties of carbon nanotubes (CNT) are affected by the adsorbents on their surface. This can be used to design gas sensors based on the change of the electrical properties of the CNT when gas molecules adsorb on their surface. CNT are p-type conductors. Therefore, oxidising and reducing gases will have different effect on the conduction properties of the CNT. Figure (a) shows a scanning electron microscopy image of the contacted macroscopic CNT rope. Figure (b) shows a magnification of the evidencing the nanoscopic character of the rope.

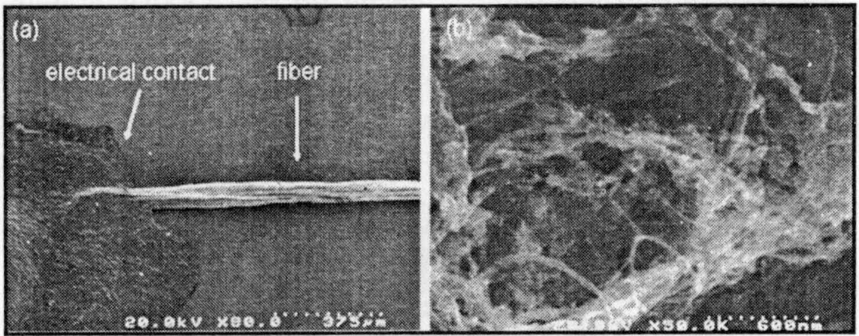

Fig. (a) SEM Image of a Contacted Macroscopic Rope. (b) Detail of the Rope showing that it is formed by CNTs.

The gas adsorption properties of these fibers are being studied in collaboration with the Department d'Enginyeria Electronica of the Universitat de Barcelona. Figure shows the room temperature response of a fibre to small quantities NO_2 and NH_3. The response is defined as the change in the resistance of the fibre when the gas is introduced in the chamber. The resistance decreases when the NO_2 is introduced since this gas removes an electron from the fibre and creates additional carriers. On the other hand, the resistance increases when the same fibre is exposed to NH_3 because this gas injects electrons to the fibre and therefore reduces the number of holes.

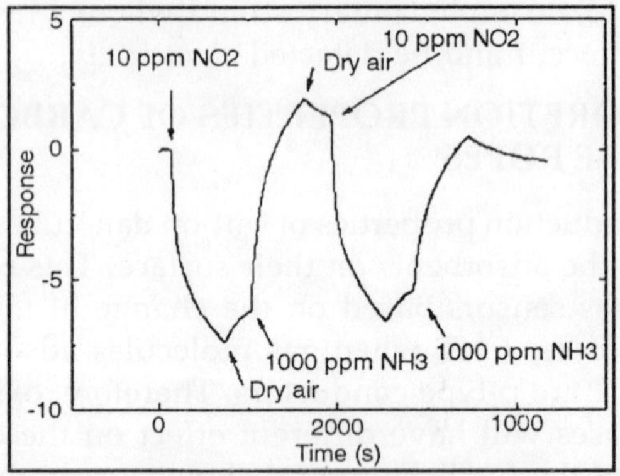

Fig. Response of a Carbon Nanotube Rope to NO 2 and NH 3. The Response is defined as the Change in Resistance when the Gas is Introduced in the Chamber.

The design of gas sensors based on these ropes is being currently under study. The main interest is to achieve selectivity by depositing catalysts that enhance the decomposition of certain gas and therefore boost the response to that certain gas. Figure shows a SEM image of a deposited catalyst particle.

Fig. SEM image of Catalysts Particle Electrodeposited

Chapter 5

Nanoscience and Nanotechnologies

The first term of reference of this study was to define what is meant by nanoscience and nanotechnology. However, as the term 'nanotechnology' encompasses such a wide range of tools, techniques and potential applications, we have found it more appropriate to refer to 'nanotechnologies'. Our definitions were developed through consultation at our workshop meeting with scientists and engineers and through comments received through the study website.

- Although there is no sharp distinction between them, in this report we differentiate between nanoscience and nanotechnologies as follows.

Nanoscience is the study of phenomena and manipulation of materials at atomic, molecular and macromolecular scales, where properties differ significantly from those at a larger scale.

Nanotechnologies are the design, characterisation, production and application of structures, devices and systems by controlling shape and size at nanometre scale.

- The prefix 'nano' is derived from the Greek word for dwarf. One nanometre (nm) is equal to one-billionth of a metre, 10–9m. A human hair is approximately 80,000nm wide, and a red blood cell approximately 7000nm wide. Figure shows the nanometre in context. Atoms are below a nanometre in size, whereas many molecules, including some proteins, range from a nanometre upwards.

- The conceptual underpinnings of nanotechnologies were first laid out in 1959 by the physicist Richard Feynman, in his lecture 'There's plenty of room at the bottom'. Feynman explored the possibility of manipulating material at the scale of individual atoms and molecules, imagining the whole of the Encyclopaedia Britannica written on the head of a pin and foreseeing the increasing ability to examine and control matter at the nanoscale.
- The term 'nanotechnology' was not used until 1974, when Norio Taniguchi, a researcher at the University of Tokyo, Japan used it to refer to the ability to engineer materials precisely at the nanometre level. The primary driving force for miniaturisation at that time came from the electronics industry, which aimed to develop tools to create smaller (and therefore faster and more complex) electronic devices on silicon chips. Indeed, at IBM in the USA a technique called electron beam lithography was used to create nanostructures and devices as small as 40–70nm in the early 1970s.
- The size range that holds so much interest is typically from 100nm down to the atomic level (approximately 0.2 nm), because it is in this range (particularly at the lower end) that materials can have different or enhanced properties compared with the same materials at a larger size. The two main reasons for this change in behaviour are an increased relative surface area, and the dominance of quantum effects. An increase in surface area (per unit mass) will result in a corresponding increase in chemical reactivity, making some nanomaterials useful as catalysts to improve the efficiency of fuel cells and batteries. As the size of matter is reduced to tens of nanometres or less, quantum effects can begin to play a role, and these can significantly change a material's optical, magnetic or electrical properties. In some cases, size-dependent properties have been exploited for

centuries. For example, gold and silver nanoparticles (particles of diameter less than 100 nm) have been used as coloured pigments in stained glass and ceramics since the 10th century AD. Depending on their size, gold particles can appear red, blue or gold in colour. The challenge for the ancient (al)chemists was to make all nanoparticles the same size (and hence the same colour), and the production of single-size nanoparticles is still a challenge today.

- At the larger end of our size range, other effects such as surface tension or 'stickiness' are important, which also affect physical and chemical properties. For liquid or gaseous environments Brownian motion, which describes the random movement of larger particles or molecules owing to their bombardment by smaller molecules and atoms, is also important. This effect makes control of individual atoms or molecules in these environments extremely difficult.
- Nanoscience is concerned with understanding these effects and their influence on the properties of material. Nanotechnologies aim to exploit these effects to create structures, devices and systems with novel properties and functions due to their size.
- In some senses, nanoscience and nanotechnologies are not new. Many chemicals and chemical processes have nanoscale features – for example, chemists have been making polymers, large molecules made up of tiny nanoscalar subunits, for many decades. Nanotechnologies have been used to create the tiny features on computer chips for the past 20 years. The natural world also contains many examples of nanoscale structures, from milk (a nanoscale colloid) to sophisticated nanosized and nanostructured proteins that control a range of biological activities, such as flexing muscles, releasing energy and repairing cells. Nanoparticles occur naturally, and have been created for thousands of years as the products of combustion and food cooking.

- However, it is only in recent years that sophisticated tools have been developed to investigate and manipulate matter at the nanoscale, which have greatly affected our understanding of the nanoscale world. A major step in this direction was the invention of the scanning tunnelling microscope (STM) in 1982, and the atomic force microscope (AFM) in 1986. These tools use nanoscale probes to image a surface with atomic resolution, and are also capable of picking up, sliding or dragging atoms or molecules around on surfaces to build rudimentary nanostructures. In a now famous experiment in 1990, Don Eigler and Erhard Schweizer at IBM moved xenon atoms around on a nickel surface to write the company logo, a laborious process which took a whole day under well-controlled conditions. The use of these tools is not restricted to engineering, but has been adopted across a range of disciplines. AFM, for example, is routinely used to study biological molecules such as proteins.
- The technique used by Eigler and Schweizer is only one in the range of ways used to manipulate and produce nanomaterials, commonly categorised as either 'top-down' or 'bottom-up'. 'Top-down' techniques involve starting with a block of material, and etching or milling it down to the desired shape, whereas 'bottomup' involves the assembly of smaller sub-units (atoms or molecules) to make a larger structure. The main challenge for top-down manufacture is the creation of increasingly small structures with sufficient accuracy, whereas for bottom-up manufacture, it is to make structures large enough, and of sufficient quality, to be of use as materials. These two methods have evolved separately and have now reached the point where the best achievable feature size for each technique is approximately the same, leading to novel hybrid ways of manufacture.

- Nanotechnologies can be regarded as genuinely interdisciplinary, and have prompted the collaboration between researchers in previously disparate areas to share knowledge, tools and techniques. An understanding of the physics and chemistry of matter and processes at the nanoscale is relevant to all scientific disciplines, from chemistry and physics to biology, engineering and medicine. Indeed, it could be argued that evolutionary developments in each of these fields towards investigating matter at increasingly small size scales has now come to be known as 'nanotechnology'.
- It will be seen that nanoscience and nanotechnologies encompass a broad and varied range of materials, tools and approaches. Apart from a characteristic size scale, it is difficult to find commonalities between them. We should not therefore expect them to have the same the same health, environmental, safety, social or ethical implications or require the same approach to regulation.

OUR SCIENCE INITIATIVES – NANOTECHNOLOGY

CCR investigators support NCI's Alliance for Nanotechnology in Cancer through the Nanobiology Programme and also work in close collaboration with the Nanotechnology Characterization Laboratory.

CCR's Nanobiology Programme (CCRNP) promotes multidisciplinary research that leads to the development of tools for nanoscale, biologically based strategies to prevent, diagnose, and treat cancer, AIDS, and biodefense-related viral diseases. The CCRNP works to understand the structure and function of biomolecules to aid in the design of nanodevices for in vivo imaging, diagnostics, and targeted drug delivery systems.

Principle investigators who are active in the programme have several areas of interest, including membrane structure and function, protein interactions, biomedical image and database analysis, structural bioinformatics, structural glycobiology, molecular information theory, and computational RNA structure.

CCRNP investigators have formed collaborative partnerships with CCR investigators in other programs, labs, and branches, including the Molecular Imaging Programme, the Radiation Oncology Branch, the Laboratory of Cell Biology, and the Laboratory of Medicinal Chemistry.

The Nanotechnology Characterization Laboratory (NCL) conducts preclinical efficacy and toxicity testing of nanoparticles intended for cancer therapeutics and diagnostics. The laboratory provides critical infrastructure support to NCI's Alliance for Nanotechnology and is a partnership between the NCI, the U.S. Food and Drug Administration, and the National Institute of Standards and Technology. The NCL assists the bionanotech community in identifying structure-activity relationships related to nanoparticle safety and efficacy.

CCR NANOBIOLOGY PROGRAMME NEWS

Computational Approaches to the Construction of RNA-Based Nanodevices

CCR computational scientists are using their multidisciplinary expertise in algorithm design, RNA structure prediction and analysis, molecular modeling and high-performance computing to develop RNA-based nanodevices. Most of the recent work on the use of biomolecules for nanodevices has concentrated on DNA and proteins, but only in a few cases has taken into account the use of RNA molecules to construct such devices. An important property of protein-free RNA systems is that human immune response is typically low or undetectable.

RNA-based systems are also very attractive for nanobiology because they are relatively easy to synthesize, and studies are finding RNA to be an equally, if not more, desirable material for designing functional nanostructures. RNA nanoconstructs can serve as building blocks and scaffolds to assemble more complex functional objects. In some cases, these shapes can self-assemble and can be of arbitrary sizes, while in other cases, they may be tailored for very specific purposes and must be within the correct size range (~20 nanometers) for the delivery of functional groups (RNA, peptides, or otherwise) to cells for controlling cell genetics, including cell death, or visualization of delivery pathways. Other potential applications include biosensors and crystallography substrates.

CCR scientists are investigating the structural properties of various RNA motifs and are working on computer applications for the interactive design of RNA nanostructures. It will be possible to visualize and manipulate three-dimensional (3D) representations of RNA interacting with other RNAs or other biomolecules. As part of this effort, an algorithm is being developed which generates nucleotide strands that form stem and loop motifs that trace a given target 3D structure. Current and future efforts involve the development of algorithms for automated sequence design, rigidity analysis, and assembly process optimization.

Novel Human Monoclonal Antibodies to Components of the IGF System

CCR researchers have developed a panel of novel human monoclonal antibodies against IGF-II, IGF-I, and IGF-IR that may potently inhibit the IGF-IR signaling function and IGF-II–mediated signaling through the insulin receptor. One of these antibodies, m610, bound with nM affinity to IGF-II and inhibited IGF-IR phosphorylation and phosphorylation of the downstream kinases Akt and MAPK as well as migration and proliferation of cancer cell lines. The researchers hypothesized that targeting IGF-II, in addition to blocking its interaction with the IGF-IR, would block that portion of the signal transduction through the insulin receptor due to its interaction with IGF-II. Lowering its level may not induce upregulation of its production as for IGF-I.

Finally, targeting a diffusible ligand, such as IGF-II, may not require penetration of the antibody inside tumors but could shift the equilibrium to IGF-II complexed with antibody so the ligand concentration would decrease in the tumor environment without the need for the antibody to penetrate the tumor. These results indicate that an immunotherapeutic potential of IgG1 m10 is likely in combination with other antibodies and anti-cancer drugs, but only further experiments in animal models and clinical trials can evaluate this possibility. These new antibodies against the IGF-IR are currently being tested for incorporation into nanoliposomes for development of multifunctional nanoparticles that can specifically target cells expressing this receptor and help to image them *in vivo*.

Selective Inactivation of Pathogenic Organisms, Viruses, and Tumor Cells for Vaccine Development

CCR researchers have discovered that labeling hydrophobic domains of viral proteins results in the inactivation of the virus without affecting its integrity or the conformation of envelope proteins.

Researchers plan to further develop hydrophobic, cross-linking probes that will have the dual effect of inactivating the virus and making the treated virus resistant to detergent solubilization. This feature will allow elimination by detergent treatment of any possible residual infectious virus from the inactivated preparation, thus making it suitable for use in prophylactic vaccination. This new technology is broadly applicable for inactivating any organism with a membrane. It may be applied to a host of pathogens as well as whole-cell cancer vaccines. CCR researchers are also exploring the applications of fluorescently labeled inactivated viruses as nanoprobes in cancer research, such as activating cross-linking/photolabeling reagents at the tumor site.

Other research foci include the design and development of lipid-based nanoparticles for antibody-mediated, tumor-specific targeting and triggered release of anti-cancer drugs or genes to various tumor tissues; the use of viral proteins to build nanofusion machines that will directly deliver their cargo to cells' cytoplasm; and the understanding of important nanoparticle-cell interactions that lead to specific recognition, imaging, and drug or gene delivery.

High-Speed Parallel Molecular Nucleic Acid Sequencing

CCR's Molecular Information Theory Group specializes in using mathematical theory to produce future-technologies, such as designing molecular machines and developing nanotechnologies.

This group has conceptualized several projects, including high-speed parallel molecular nucleic acid sequencing. Using this technique, the sequence of a single molecule of DNA or RNA can be read using a microscope to observe changes in the fluorescence of individual nucleic acid molecules being read by a DNA or RNA polymerase.

The data are collected in a computer in parallel, allowing many sequences to be determined from a tiny volume. The technique could be used to read the sequences of mRNA directly from cells, bypassing current microarray technology. It may also be fast enough to allow individual laboratories to read entire genomes with a single device. This nanotechnology would have diagnostic uses, such as rapid identification of mutations that cause cancer and other genetic diseases. The Molecular Information Theory Group is also developing other future-technologies, such as a molecular computer and a molecular rotation engine.

MUTANT GLYCOSYLTRANSFERASES ASSIST IN THE ASSEMBLY OF GLYCOCONJUGATES

CCR researchers have designed novel glycosyltransferases, based on their structural information, that have broader or requisite donor and acceptor specificities. Several mutant glycosyltransferases have been generated that can transfer a sugar residue with a chemically reactive functional group to N-acetylglucosamine, galactose, and xylose residues of glycoproteins, glycolipids, and proteoglycans (glycoconjugates). These glycoconjugates are cross-linked via modified glycan moieties and are assisting in the assembly of bionanoparticles that are useful for the development of the targeted-drug delivery system and contrast agents for MRI. The reengineered recombinant glycosyltransferases are also making it possible to synthesize oligosaccharides for vaccine development and remodel the oligosaccharide chains of glycoprotein drugs.

Self-Assembly of Biological Building Block Motifs for Nanoscale Cancer Biology Applications

CCR computational biologists are applying their expertise in the principles of molecular structures and the mechanisms of their formation to nanobiology. They employ biological (particularly protein) building block motifs in a self-assembly process to engineer molecular-sized components for particular functional designs. The key is to be able to control and manipulate the self-assembly of the building blocks to obtain diverse shapes, sizes, chemistry of surfaces, and dynamics. The database of protein structures is populated by a vast collection of protein and building block geometries and chemical properties.

Their synthesis is fast, and they are expected to be nontoxic. Hence, they are very attractive as starting points for nanodesign. The challenge is to develop predictive, mechanistic schemes that allow controlling the assembly of the intermediate conformational states. The strategy is to choose potential building blocks and enhance their population times through introduction of native and non-native residues. The building blocks are then assembled, and their stabilities are estimated through high-performance computing. Proteins are also used in strategies that mimic naturally occurring functional constructs. It is expected that such advanced computational approaches will lead to a considerable acceleration of nanodesign.

COLLABORATIVE NEWS

Bridging the Imaging Gap in Nanobiology with 3D Electron Microscopy

Emerging methods in 3D biological electron microscopy provide powerful tools and great promise to bridge a critical gap in imaging in the biomedical size spectrum. This gap comprises a size range of great interest in biology and medicine that includes cellular protein machines, giant protein and nucleic acid assemblies, small subcellular organelles, and small bacteria. These objects are generally too large or too heterogeneous to be investigated by high-resolution X-ray and NMR methods, but the level of detail afforded by conventional light and electron microscopy is often not adequate to describe their structures at resolutions high enough to be useful in understanding the chemical basis of biological function.

CCR investigators are using electron microscopic imaging to discover and analyse biological complexity within the size gap with linear dimensions of about 50–1000 nm. Ultimately, the understanding of cellular architecture gained at this level will be crucial in designing effective strategies for disease prevention and treatment. A key mission is to quantitatively describe the spatial and temporal architecture of key molecular machines that fall into this "nano gap". Areas of current interest include:

- The development and application of novel technologies for 3D electron microscopy of specimens ranging in size from small molecules to tissues, including automated approaches to analyse the molecular structure and sub-cellular location of a variety of nanoparticles,
- Determination of the dynamic spatial and temporal architectures of cellular structures and molecular machines involved in fundamental processes such as energy transduction, cell division, and chemotaxis,
- Determination of molecular mechanisms underlying the neutralization and cellular entry of HIV.

Development and Characterization of Affibody-Based Bioconjugates for Molecular Imaging

CCR researchers are developing an innovative strategy for individualized treatment of HER2-positive cancers by combining a non-invasive method for monitoring of HER2 in vivo and HER2-specific delivery of therapeutic agents. Affibody molecules obtained from our CRADA partner in Sweden are used as the targeting agent. These very stable and highly soluble a-helical proteins are relatively small (8.3 kDa) and can be readily expressed in bacterial systems or produced by peptide synthesis. The His6-Zher2:324 binds to HER2 receptors with high affinity (22 pM) and is available with cysteine at the carboxy-terminal to facilitate conjugation.

For imaging purposes, these molecules will be labeled with radionuclides. For therapy, the His6-Zher2:324 will be conjugated with thermo- or radio-sensitive liposomes when labeled with beacons for in vivo imaging and loaded with therapeutic agents (e.g., toxins, radiosensitizers, or kinase inhibitors) that will allow local drug release defined by real-time monitoring of their distribution. These nanoparticles will provide means for HER2-specific delivery of a variety of tumoricidal agents, including those whose application is currently limited due to their hydrophobicity and that, thereby, complement current therapeutic strategies. This approach, involving assessment of target presence and distribution in an individual patient followed by optimized, target-specific drug delivery, should significantly improve the efficacy of cancer treatment while reducing side effects.

Immunotherapeutic Cocktail of Nanoparticles to Attack Tumor Cells

Researchers in the Laboratory of Medicinal Chemistry (LMC) synthesize gold nanoparticles coated with tumor-associated carbohydrate antigens (TACAs)—in particular, the Thomsen-Friedenreich (TF) antigen disaccharide. This sugar is present on over 90% of carcinoma cells but rarely displayed on normal tissue. Researchers have shown that the carbohydrate was functional on the particle and that, when injected into mice with implanted breast tumors, these particles inhibited metastasis to the lungs. In addition, researchers are preparing nanoparticles that are coated with glycopeptides that contain the TF antigen as precursors to a possible anti-cancer vaccine. The goal is to prepare multifunctional particles as part of a cocktail for immunotherapy targeted against tumor cell surface carbohydrate epitopes and their surrounding environment.

Researchers also have used CCR's synthetic precursors to prepare the first quantum dots coated with the TF TACA. Quantum dots are highly useful semiconductor nanocrystals that have unique photoluminescent properties.

Researchers were successful in accomplishing the first de novo synthesis of carbohydrate-coated quantum dots. From there, they proceeded to enhance their luminescent properties by making hybrid particles with small organic acids.

They were able to synthesize very robust particles with high quantum yields and biofunctional sugars on the surface. Researchers were able to selectively label lung metastasis cells with these particles and are proceeding to tackle many other cell-labeling applications with the LMC's newly developed protocol for quantum dot synthesis.

Labeled Nanoparticles Reveal Tumor Details

Conventional magnetic resonance imaging (MRI) contrast agents or dyes are widely used in hospitals today, but they have a number of limitations.

Their small molecular size means that they can easily escape from vessels and increase signal or "enhance" non-specifically in the body. They are also notoriously difficult to target to specific tissues, including cancerous tissues. NCI researchers have recently designed a new nano-sized MRI contrast agent based on a family of macromolecules known as dendrimers.

These nanoparticles are composed of branching molecules (dendron in Greek means tree) that allow multiple atoms of gadolinium, the paramagnetic agent that enhances MRI, to attach to tissues.

The slower molecular tumbling rate of nanoparticles enhances their relaxivity and, therefore, their effect on the image.

In addition, the large number of potential binding sites on the surface of dendrimers allows them to be dual or triple labeled with other imaging probes, such as radionuclide or optical probes, as well as with ligands, which target surface receptors found mainly on cancer cells.

NANOTECHNOLOGY CHARACTERIZATION LABORATORY NEWS

NANOPARTICLES CHARACTERIZED

The Nanotechnology Characterization Laboratory (NCL) is now actively characterizing nanoparticles intended for clinical applications.

At the close of fiscal year 2005, the NCL had accepted 26 particles for characterization, exceeding the 24 particles forecasted in the NCL business plan. Efforts are now focused on continuing to solicit nanoparticles, subjecting the particles to the NCL's three-phase assay cascade, and identifying parameters that influence biocompatibility.

Chapter 6

Impact of Nanotechnology

ORAL INHALATION TECHNOLOGY

Regulatory awareness has led to the need for new approaches when regulating nanotechnology in the inhaled area. Opportunities as well as challenges exist. If addressed, they can open the doors for new IP to be generated along with considerations that can be incorporated in Life Cycle Management.

The FDA announced in August 2006 that it will form an internal FDA Nanotechnology Task Force. The group responsibilities will be to establish regulatory approaches in the development of safe and effective FDA-regulated products that use nanotechnology, affecting virtually every product category that FDA regulates—from pharmaceuticals and devices to cosmetics and food supplements. In July 2007 the FDA released a report that recommends the agency to consider guidance and steps to address benefits and risks of drugs and medical devices using nanotechnology.

Eschenbach (MD), a Food and Drug commissioner stated that "Nanotechnology holds enormous potential for use in a vast array of products".

Guidance would clarify what information needs to be supplied to the FDA when developing nanotechnology-based products and, as the uncertain nature of nanotechnology and the potential rapid development clearly highlights the need for predictable, transparent and consistent regulatory pathways.

A need to identify and access data needs in the FDA regulatory process of nanotechnology and products and will cover both interactions and possible biological effects of nano-sized materials.

The report also flags for a commitment to a more transparent public process when FDA develops its regulatory policies around nanotechnology. Public participation will ultimately win public confidence and be paramount when developing good policies in the agency's oversight of nanotechnology and nanotech products.

Many of the products that incorporate nanotechnology are non-pharmaceuticals comprising materials and line extensions of these.

In the next five years it is foreseen that we will see products in the field of solar energy, batteries, displays and e-paper, nanotubes and nanoparticle composites, catalysts, coatings, paints, alloys, insulation, filters, glues, abrasives, lubricants etc. The time to market these products varies and depends both on the technological complexity as well as the regulatory complexity and product life cycle as can be seen in Figure.

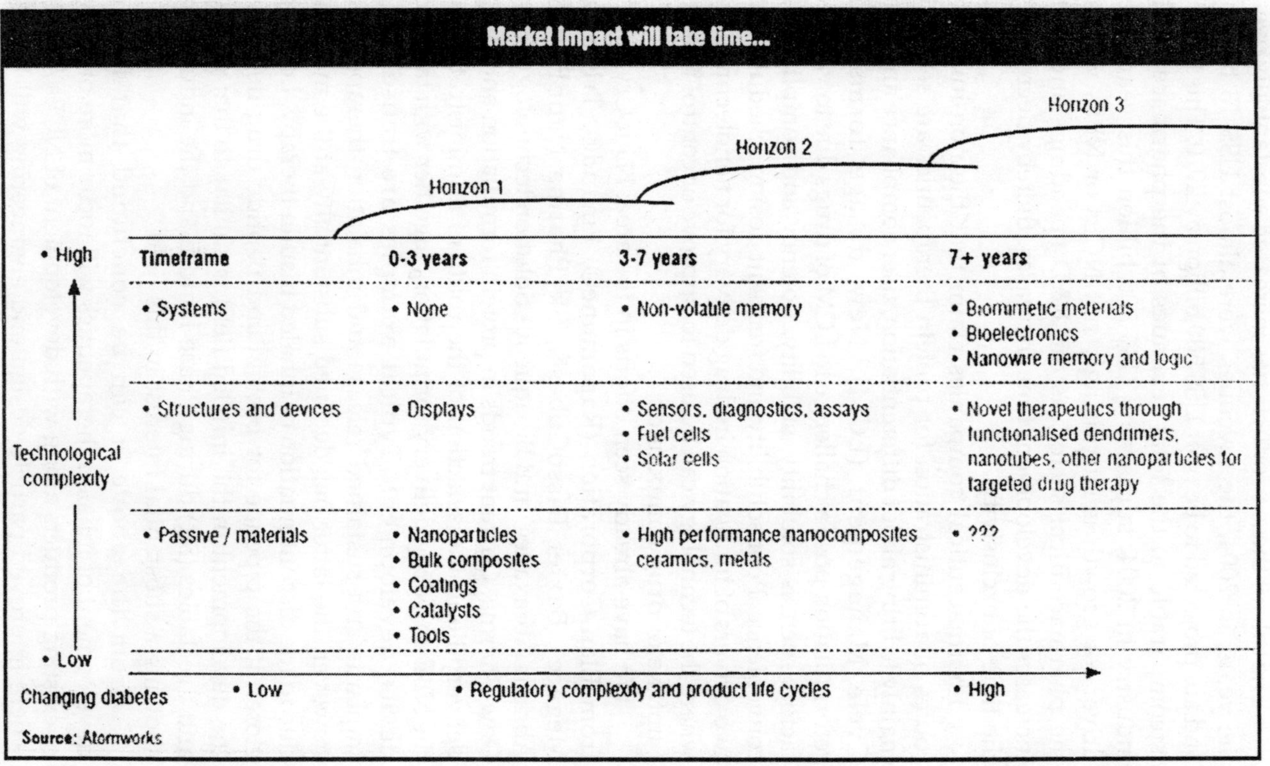
Market Impact will take time...
Horizon 1
Horizon 2
Horizon 3
Timeframe
0-3 years
3-7 years
7+ years
• Systems
• None
• Non-volatile memory
• Biomimetic materials
• Bioelectronics
• Nanowire memory and logic
• Structures and devices
• Displays
• Sensors, diagnostics, assays
• Fuel cells
• Solar cells
• Novel therapeutics through functionalised dendrimers, nanotubes, other nanoparticles for targeted drug therapy
• Passive / materials
• Nanoparticles
• Bulk composites
• Coatings
• Catalysts
• Tools
• High performance nanocomposites ceramics, metals
• ???
• High
Technological complexity
• Low
Changing diabetes
• Low
• Regulatory complexity and product life cycles
• High
Source: Atomworks

The sale of these products has increased substantially over the years; in 2004, the revenues were almost US$ 13 billion and are projected to be over US$ 500 billion by 2010. This figure is comparable to the total revenues of the pharmaceutical industry in 2004, surpassing US$ 500 billion this year. The driver for growth is surprisingly not NCEs or NBEs within the pharmaceutical sector, as the R&D spendings and cost increases the development time increases, thereby decreasing the time of exclusivity.

Pharmaceutical companies claim that the non-invasive routes of administration for protein therapeutics are selected mainly to bring about differentiation, user compliance and Life Cycle Management (LCM). New dosage forms and reformulation pose a challenge to LCM of drugs due to various factors such as solubility, stability, potency and compliance / convenience. Poor solubility and frequent dosing schedules are two drivers of innovation in drug delivery for nanotechnology where the technology can be used to improve or control release / uptake of drug compounds.

We have already seen efforts in the area: NanoCrystals® from Elan Corporation (Rapamune®, Emand®, TriCor®, Megace), Baxter DissoCubes®, SkyPharma NanoEdge® Abraxis Abraxane® and Bioaqueous Solution technology from Dow Pharma. Current trends in protein formulation and drug delivery lie in the selection of the route of administration of the NBE and drug delivery formulation, together with tailored device development. Typical examples are transdermal implants and sustained release and inhalation therapeutics; however, the latter has declined substantially at the moment due to the discontinuation of inhaled insulin therapy. Looking closer at the pipeline for protein and peptide drug delivery, the development in the inhaled field is similar to that of oral and injectables (SR) during Phase I and II and the industry is showing a substantial interest in this field.

From the above it can be concluded that future development of inhaled therapeutics will focus more on LCM of existing products along with development of NCEs / NBEs where improved particle design and engineering will be key development trends. The size, shape and composition are the main factors of interest to exploit further. Both the size and the shape factors such as structure, form and topology will have an influence on the drug delivery from the device as the binding or connecting forces can be moderated to improve small or nanoparticle dispersion during delivery.

Classically, drug delivery via inhalation has predominately centred on micron particles as these have been able to deliver defined metered dose during one or more inhalations. Switching to submicron or nanoparticles, the number of particles easily increases by a factor of thousand or million for the same dosing regime and the adhesive / cohesive forces multiply to the same extent. Exploiting options around shape and composition is a must to minimise the effect of cohesion / adhesion in drug delivery. The composition, shape, density and the total number of particles (e.g. the total surface coating of the lung) may have an effect on the dissolution rate, uptake and biodegradability of the particles when they reach the lung as compared to larger micron sized particles.

This also reflects the possibilities of using nanoparticles and nanomaterials in potential rapid topical drug delivery. Nanoparticles can also be a catalyst in the development of new carrier molecules and materials such as nanoshells and assembled structures. All of these features can of course be incorporated into the future inhalation therapy.

The inhalers that are offered today are divided into three segments—Dry Powder Inhalers (DPI), pressurised Metered Dose Inhalers (pMDI) and nebulisers or Soft Mist Inhalers (SMI). In DPI, the drug is formulated as a dry powder, containing the active drug component and excipient or carriers. A variety of DPIs are available today and they can have either a reservoir of powder or be supplied with a number of unit blisters / capsules for individual dosing.

The pMDI and the SMI (nebuliser) use a dry powder formulation but the powder is suspended in a Hydro Fluoro Alkanes (HFA) or Chloro Fluoro Carbons (CFC) propellant in the case of pMDIs and in water in the case of a nebuliser or SMI. A pMDI is a reservoir system and often holds 60 to 120 doses in the reservoir. The SMI is a reservoir system containing multiple highly concentrated μl doses (typically10 to 50 μl) whereas the nebuliser uses a lower concentrated ml dose (typically 1 to 3 ml). Stability and degradation is an issue and is more accentuated in the water-based nebuliser / SMI systems than in pMDI and DPI systems due to the better chemical stability in non-aqueous systems.

Inhaler devices can of course be used for delivering nanoparticles, both as a drug and as a carrier system for various purposes.

Care has to be taken as the implementation of nanoparticles in inhalation therapy poses some major challenges to the developer.

Due to the nature of nanoparticles, they will give rise to many formulation issues when blended with excipients as the surface to area ratio is large compared to micron-sized particles and thus will have an impact on blending and aggregation properties of the formulation. For liquid (aqueous and HFA systems) aggregation resulting in flocculation and caking together with risk of adhesion to walls and device mechanism may have an impact on device dosing properties (uniformity and over time).

Dispersion of the particles can also pose some challenges, especially for DPIs where the patient inhalation effort drives the de-aggregation of particles, a factor that is known to vary substantially depending on the disease and state of the patient. The residue in the blister / capsule and device will also be dependent on the nature of the formulatin used and opens for both issues and possibilities when employing nanotechnology (particles and structures) in inhalation-based delivery of drugs.

After solving these issues, there may still be other issues to address when looking at PK / PD profiles. This is especially critical of generic applications where equivalent PK / PD profiles and dosing to marketed products are essential. The difference in particle size and surface area coverage along with possible differences in dissolution and biodegradability / availability can have a major impact and must be addressed early in the development.

Example Formulation : CFC to HRA Conversion

Gamma scintigraphic images of HFA (QVAR) and CFC beclomethasone pMDl

Region Device	HFA BDP 40 µg	CFC BDP 42 µg
Oral	31%	94%
Lung	51%	0.4%
Exhaled	18%	01%

Observe: Both particle and plume velocity dependent

A wide range of possibilities in which nanoparticles can be used in inhalation therapy is felt since long. When IVAX and 3M developed the QVAR pMDI system as a CFC to HFA conversion of an existing CFC product containing a micron-sized suspension, the resulting HFA formulation became a solution and not a suspension. When activating the QVAR® pMDI, the HFA boils off resulting in solid particles whose size depended on the drug concentration in the HFA formulation and the initial droplet size of the HFA droplet distribution.

The resulting solid particles were much smaller than the suspended particles in the CFC suspension formulation and were more likely to penetrate into the lung as compared to the CFC suspension particles. The reported results can be seen in Table which clearly shows an enhanced lung deposition of the QVAR® as compared to the conventional CFC product on the market. This clearly highlights opportunities for nanoparticle applications. But, it also highlights the differences in dosing and possible PK / PD issues for generica or line extension (LCM) applications.

The success of nanoparticle applications in the inhaled field depends on the feasibility of designing nanoparticle formulations that are stable and can be delivered successfully and repeatedly to the patient. Many issues need to be considered and thoroughly tested in order to achieve the goals set in the development phase. If this is successfully done, new options securing IP-enabling novel NCE / NBE or even line extension applications may be thought of. At the moment, this area is fairly unexploited generating new IP in the field giving a cutting edge to novel products in the future for those pharmaceutical companies working in the inhaled product segment.

GLOBAL NANOTECHNOLOGY RESEARCH

In 2003–05, a comprehensive text-mining study was performed to survey the technical structure and infrastructure of the global nanotechnology research literature, as well as the seminal nanotechnology literature2,3.

Based on the wide-scale interest generated by these reports, it was decided to update and expand the study using more recent data, a much more comprehensive query and more sophisticated analytical tools.

In the updated study, text mining was used to extract technical intelligence from the open source global nanotechnology and nanoscience research literature (SCI/SSCI databases). The following were identified:

- The nanotechnology/ nanoscience research literature infrastructure (prolific authors, key journals/ institutions/countries, most cited authors/journals/ documents);
- The technical structure (pervasive technical thrusts and their inter-relationships);
- Nanotechnology instruments and their relationships;
- Potential nanotechnology applications;
- Potential health impacts and applications,
- Seminal nanotechnology literature.

The results are summarized in this article. A more detailed report on the results and methodologies of this updated study can be found in Kostoff *et al*. This article is an overview of the highlights of the total study, including the production efficiency of seminal nanotechnology documents.

The results are divided into four main sections: Infrastructure, Technical structure, Instrumentation and Applications. The Applications section is further divided into non-medical and medical. The results will be presented in the order listed above.

Next, the seminal nanotechnology literature production

efficiency will be presented. Infrastructure describes the performers of nanoscience/ nanotechnology research at different levels, ranging from individual to national performers, and it includes archived literature as well.

Technical structure identifies the pervasive technical thrusts (and their inter-relationships) of the nanoscience/nanotechnology literature.

Instrumentation provides both infra-structure and technical structure of the subset of the nanoscience/ nanotechnology literature that addresses specific instruments. Applications provides the infrastructure and taxonomy of the subset of the nanoscience/ nanotechnology literature that addresses specific non-medical and medical applications.

APPROACH

An extensive nanotechnology/nanoscience-focused query (300 + terms) was applied to the SCI/SSCI database. The nanotechnology/nanoscience research literature technical structure (taxonomy) was obtained using computational linguistics, especially document clustering. The nanotechnology/nanoscience research literature infrastructure (prolific authors, key journals/institutions/countries, most cited authors/journals/documents) for each of the clusters generated by the document clustering algorithm was obtained using bibliometrics. The instrumentation literature associated with nano-science and nanotechnology research was examined. About 65,000 nanotechnology records for 2005 were retrieved from the SCI/SSCI, and ~27,000 of these were identified as instrumentation-related.

All the diverse instruments were identified and their associated documents categorized in a hierarchical taxonomy. Metrics associated with research literature for specific instruments/instrument groups were generated. The applications literature associated with nanoscience and nanotechnology research was examined. Through visual inspection of 60,000 of the abstract phrases of the same downloaded 2005 records, all the diverse non-medical applications were identified and their associated documents categorized in a hierarchical taxonomy. Metrics associated with research literature for specific applications/applications groups were generated. For medical applications, a fuzzy clustering algorithm (where a record could be assigned to multiple clusters) was applied to the downloaded 2005 records. A sub-network that encompassed all the medical applications was identified. Again, metrics associated with research literature for specific medical applications were generated.

RESULTS

INFRASTRUCTURE

Country Publications:

- Global nanotechnology research article production exhibited exponential growth for more than a decade.
- The most rapid growth over that time period came from East Asian nations, notably China and South Korea.

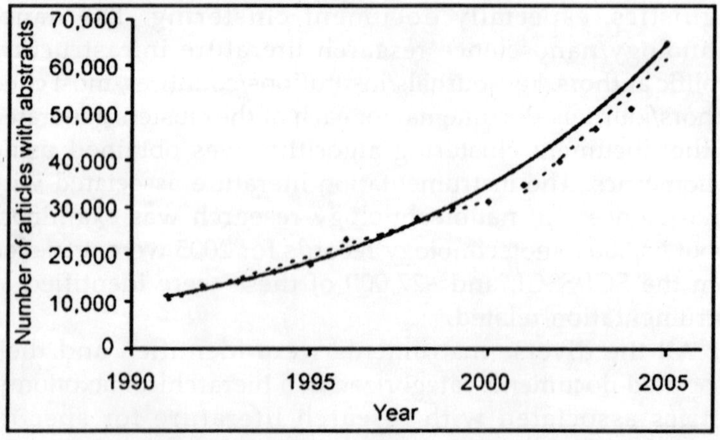

Fig. SCI/SSCI Articles vs Time: Total Records Retrieved

- Some of this apparent rapid growth (in China, for example) is partially due to
- A country's researchers publishing a non-negligible fraction of total papers in domestic low impact factor

journals.

- These journals being accessed recently by the SCI/SSCI, rather than due to growth based on increased sponsorship or productivity.
- China's representation in high impact factor journals was small, but increasing.
- From 1998 to 2002, China's ratio of high impact nanotechnology papers to total nanotechnology papers doubled, placing the country at parity for this metric with the advanced nations of Japan, Italy and Spain.
- The US remained the leader in aggregate nanotechnology research article production.
- In some selected nanotechnology sub-areas, China had achieved parity or taken the lead.

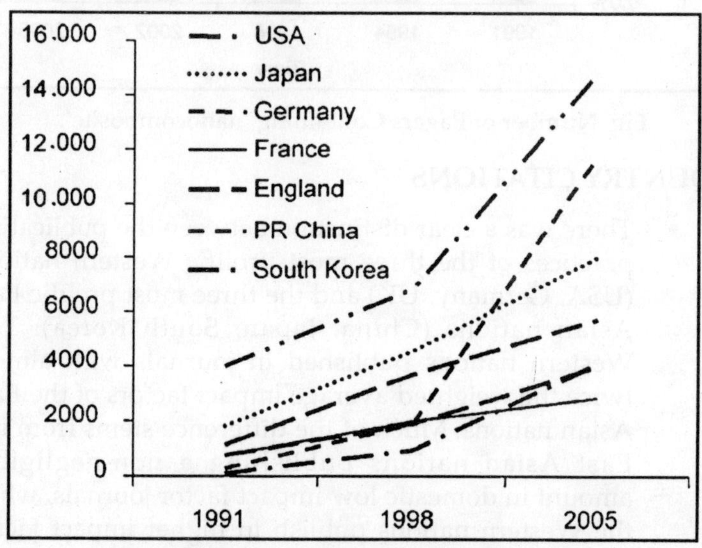

Fig. Country Comparison Time Trend (Number of Articles vs Time).

- South Korea started even further behind China in both total nanotechnology publications and highly cited papers, but has advanced rapidly to become a secondtier contender in total and highly cited papers.

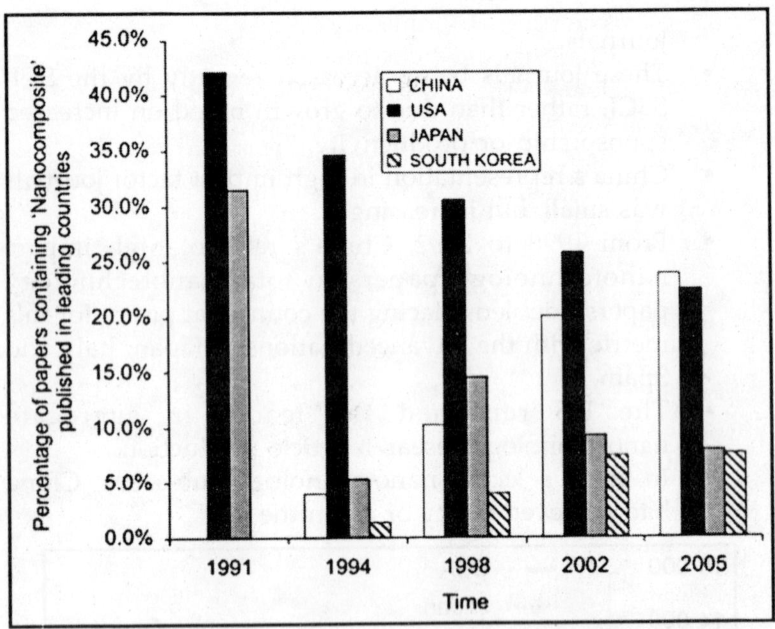

Fig. Number of Pagers Containing 'nanocomposite'.

COUNTRY CITATIONS

- There was a clear distinction between the publication practices of the three most prolific Western nations (USA, Germany, UK) and the three most prolific East Asian nations (China, Japan, South Korea). The Western nations published in journals with almost twice the weighted average impact factors of the East Asian nations. Much of the difference stems from the East Asian nations publishing a non-negligible amount in domestic low impact factor journals, while the Western nations publish in higher impact factor international journals.
- Two countries that led in production of the most cited nanotechnology papers were the US (126) and Germany (31). They accounted for 40% of the most cited nanotechnology papers.
- The high paper volume production East Asian

countries of China and South Korea accounted for 2% of the most cited nanotechnology papers.

- Despite the increased paper productivity from East Asian countries, the US continued to generate the most cited nanotechnology papers.

TECHNICAL STRUCTURE

The total retrieved nanotechnology database for 2005 was examined from four perspectives to identify pervasive thematic thrusts: document clustering, autocorrelation mapping, factor analysis and cross-correlation mapping. Each perspective provided valuable insights on the fundamental nanotechnology literature structure. Only document clustering results are presented here.

Document Clustering

The database was divided into 256 thematic clusters by the clustering algorithm. USA produced most papers in 169 thrusts, China led in 70, Japan led in 15, and India, South Korea and Spain each led in one. A hierarchical taxonomy was constructed from these 256 elemental clusters. Of the sixteen fourth-level categories in taxonomy, China was the publication leader in six. Specifically, China led in: Properties of thin films; Diamond films; Applications of carbon nanotubes; Multiwalled nanotubes; Nanomaterials and nanoparticles, and Polymers, composites and metal complexes (categories with solid shading denote publication lead by China, and those with vertical lines and shading denote publication lead by Japan.

Light shading means category leader has 100–125% of the USA publications; medium shading 125–150%; dark shading >150%). Essentially, China led in the materials and nanostructures component of the database, whereas USA led in the physical science phenomena and biomedical components.

INSTRUMENTATION

A wide variety of instruments are used in nanoscience and nanotechnology research. Key among these are X-ray diffraction

Table: Four-level Hierarchical Taxonomy

LEVEL 1	LEVEL 2	LEVEL 3	LEVEL 4
Quantum phenomena, Optics, Electronics, Magnetism, Tribology, and Films (32,983 records)	Quantum phenomena, Optics, Electronics, Magnetism, and Tribology (26,077 records)	Quantum phenomena (3326 records)	Quantum dots (2028 records)
			Quantum wells, Wires, and States (1298 records)
		Optics, Electronics, Magnetism, and Tribology (22,751 records)	Optics and Electronics (16,432 records)
			Magnetism and Tribology (6319 records)
	Films (6906 records)	Thin films (4760 records)	Properties of thin films (2251 records)
			Applications of thin films (2509 records)
		Deposition of films (2146 records)	Deposition of thin films (1752 records)
			Diamond films (394 records)
Nanotubes, Nanomaterials, Nanoparticles, Polymers, Composites, Metal complexes, and Bionanotechnology (31,742 records)	Nanotubes (3211 records)	Multi-walled nanotubes (2350 records)	Applications of carbon nanotubes (474 records)
			Multi-walled nanotubes (1876 records)
		Single-walled nanotubes (861 records)	Single- and double-walled nanotubes (447 records)
			Single-walled nanotubes (414 records)
	Nanomaterials, Nanoparticles, Polymers, Composites, Metal complexes, and Bionanotechnology (28,531 records)	Nanomaterials, Nanoparticles, Polymers, Composites, and Metal complexes (22,686 records)	Nanomaterials and Nanoparticles (14,263 records)
			Polymers, Composites, and Metal Complexes (8423 records)
		Bionanotechnology (5845 records)	DNA (775 records)
			Proteins and Cellular components (5070 records)

(XRD), electron microscope variants, atomic force microscopy, scanning tunnelling microscopy and spectroscopy variants.

Instrument Taxonomy

Hierarchical taxonomy offered the following insights:

- In this nanotechnology instrumentation study, China produced about 25% more papers than the USA (shading represents China's publication leadership; darker shading represents stronger publication leadership). By contrast, in the full nanotechnology study, USA produced about 25% more papers than China.
- Much of China's over-production occurred in the XRDrelated categories, but there was some over-production in transmission electron microscopy and NMR and calorimetry-related categories as well.
- The US dominance was in atomic force microscopy.
- Because of the large Chinese and South Korean contributions to the nanotechnology instrumentation literature, author-name analysis at aggregate levels was not effective; Asian names are usually monosyllable, many times with no middle names. Due to the relatively high frequency of paper publications, there is good possibility that the same last name represents multiple authors. Potential name disambiguation is under study.
- Even though USA has a large presence overall, relatively few US institutions were listed among the most prolific in the nanotechnology instrumentation papers. The Asian and European efforts appeared concentrated in relatively few but large institutions.

APPLICATIONS

The study also identified the main nanotechnology applications, both medical and non-medical, as well as the related science and infrastructure. These relationships will allow the potential user-communities to become involved with the applications-related science and performers at the earliest stages, to help guide the science conversion towards specific user needs most efficiently.

Non-medical applications: Applications thrust areas – Factor analysis. Factor analyses were performed to show the thematic areas in non-medical applications. A six-factor analysis showed the following themes:

- Factor 1: Optoelectronics
- Factor 2: Tribology
- Factor 3: Lithography
- Factor 4: Control systems
- Factor 5: Devices
- Factor 6: Microsystems.

Applications thrust areas – Factor analysis and visual inspection. The main non-medical applications thrust areas identified above were augmented by important but non-networked thrusts, and the nine resulting themes were related to science and infrastructure by co-occurrence matrices. Also, the total non-medical applications was combined into one unit, and related to science and infrastructure by cooccurrence matrices. For non-medical applications:

- USA led in total non-medical applications publications and in six out of nine themes in high-tech research areas such as devices, sensors and lithography. China led in publications in three traditional areas: catalysis, tribology and electrochemistry.
- In total non-medical applications, two of the top three institutions were Chinese. However, USA was well represented by the large State University systems of the University of California and University of Illinois.
- The journal *Applied Physics Letters* appeared in the top layer in seven of the nine themes and was by far the leader in total non-medical applications publications. *Journal of Physical Chemistry B* appeared in four of the nine themes, as also *Journal of Applied Physics. Medical applications:* Applications thrust areas – Visual inspection/fuzzy clustering. A medical applications categorization constructed from visual

inspection of the detailed fuzzy clustering categories showed five broad thematic categories:

- Cancer treatment
- Sensing and detection
- Cells
- Proteins
- DNA.

Applications thrust areas – Fuzzy clustering. For medical applications, analysis of nineteen thematic categories obtained from fuzzy clustering of the total 2005 nanotechnology database revealed the following:

- USA was the publication leader in total health types, and in all the thematic areas as well, mostly by a wide margin. China was the second most prolific in seven thematic areas, Japan in six, Germany in four and England in two.
- The University of California system led in five clusters, the Chinese Academy of Science led in four, and the National University of Singapore led in three. The University of California and the Chinese Academy of Science were the most prolific in the non-medical applications as well, but their orders were reversed. The National University of Singapore was a prolific contributor, especially in pharmaceuticals and biomaterials.
- The journal *Langmuir* contained the most nanotechnology articles in total health, and was in the top layer of ten of nineteen themes. The only journals in common in the top layers of applications and health were *Langmuir* and *Journal of Physical Chemistry B*.

Production Efficiency of Global Nanotechnology Literature

The global nanotechnology research literature has two main components: spatial and temporal. The spatial component covers present-day nanotechnology research being conducted globally. The temporal component reflects the impact that vintage literature has had on modern-day nanotechnology research. Both the temporal and spatial components need to be understood for full comprehension of global nanotechnology research, and for the establishment of strategic nanotechnology policy.

Assessment tools and processes have advanced sufficiently to allow an integrated picture of nanotechnology to be obtained.

The summary material presented earlier concentrates on the spatial component. The remainder of this article will concentrate on one aspect of temporal component, production efficiency of the seminal nanotechnology literature.

All the nanotechnology documents published between 1991 and 2005 were downloaded. Then, the subset with the highest number of citations was extracted, and a text mining analysis of that subset was performed to obtain the characteristics of the most cited nanotechnology documents4. Following this, the relationship between document production and seminal paper production for countries was identified.

Relation of seminal nanotechnology document production to total nanotechnology document production: There is a substantial value in understanding the efficiency of seminal nanotechnology document production, i.e. the ratio of seminal nanotechnology documents produced to over-all nanotechnology documents produced.

The present short section addresses some methods for arriving at this ratio. Citations (and publications) for nanotechnology documents published in two specific years were examined.

The purpose was to obtain some time trend data as well as better statistics than one year's data could provide. All nanotechnology documents for 1998 and 2002 were retrieved and analysed. These years were selected to be as close to the present as possible, in order to insure currency of findings, yet sufficiently vintaged to insure accumulation of adequate citations.

Normalized country production of seminal nanotechnology papers: The main nanotechnology query in this study4 was used to retrieve documents from the SCI/SSCI for 1998 and 2002.

Distribution of number of publications among institutions and countries was generated using the Analyse function of the SCI search engine. Then, the publications for each year were ordered according to Time cited. The most highly cited publications were extracted, and the country and institution distributions for those documents were generated.

The country and institution publication distributions were then compared to the citation distributions. This allowed identification of countries whose citation fractions were greater than their publication fractions (and thus were producing highly cited papers more efficiently than their publication statistics would predict).

Chapter 7

PC-TV Integration

With recent advances in digital hardware and network technologies, personal computers (PC) have become very popular for home computing and entertainment. The primary household entertainment medium is still the television (TV). The current analog television standards were introduced in the 1950's, and there is not much scope for improvement in audio and video quality. High definition television (HDTV) has recently been introduced in the home entertainment field to provide superior quality video and audio through the use of digital technology.

The HDTV can be considered as a step towards PC-TV integration. Note that the HDTV signal is completely digital, and is incompatible with the existing TV standards (i.e., NTSC, PAL and SECAM). Hence, as an intermediate step, the digital HDTV signal can be converted to an analog signal using a set-top box that contains an MPEG-2 decoder to reconstruct the video from the compressed bit stream. The output of the set-top box can be displayed on an analog TV. Set-top boxes are now appearing with an Internet browser and programming storage options. The HDTV will typically have a high resolution monitor, and contain advanced digital circuitry. It is expected that this standard will pave the way for future PC-TV integration.

PC-TV CONVERGENCE

The concept of PC-TV integration has been around for several years. Miyahara *et al.* have proposed a scheme to use an NTSC television set as a computer terminal with built-in

Internet and personal computing functions. A block diagram of the scheme is shown in Figure. The architecture includes a complete motherboard, video interface card, hard drive and CD-ROM. The CD-ROM is used for multimedia applications such as music and online encyclopedias. However, the architecture is neither flexible, nor designed for remote computing, and the cost of implementation is high.

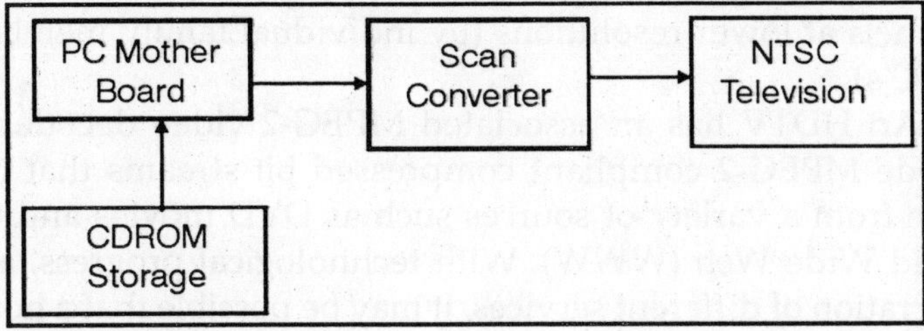

Fig. Architecture of PC-TV Integration

Other options involve connecting the computer to a television set. Computer video cards with NTSC output circuitry have been used for computer game applications. However the bandwidth of the connection is very high and not suitable for remote computing. Due to the rapid growth in communication technology, network computers (also known as disk-less computers) have become feasible. In these computers, local storage can be eliminated in favour of a remote server. However, the cost of the network computers is high, and is comparable to a traditional personal computer.

MOTIVATION

TV's are becoming increasingly portable with the introduction of the Internet and Internet based broadcasting (such as windows media format television). Two types of media channels for the consumer are expected to be seen in the future; a low resolution, low bandwidth portable channel, and a high bandwidth high resolution channel. The low resolution low bandwidth channels can be used for applications such as portable PC based TV on demand.

On the other hand, the high bandwidth, high resolution channels can be used for a home theatre based setup in a living room (at homes) where the HDTV programming using projection televisions at regularly scheduled time slots (satellite broadcasting) will take place without the use of personal computers doing the decoding.

The elimination of HDTV decoding by the personal computer was due to the high bandwidths and the coding complexity required. This would restrict home theatre to family viewing versus personal viewing of special interest channels at lower resolutions (by individual family members on PC's).

An HDTV has an associated MPEG-2 video decoder to decode MPEG-2 compliant compressed bit streams that can come from a variety of sources such as DVD movies and the World Wide Web (WWW). With technological progress, and integration of different services, it may be possible that a home computing facility will be provided as a utility (similar to the cable services) in the near future.

The high-speed computer server would be situated in a remote location, which can serve a large number of users. The integration of television and the personal computer will provide an additional motivating factor for HDTV replacement of NTSC televisions as a separate personal computer would no longer be required for most households.

OBJECTIVE

In this report we develop an efficient architecture to integrate personal computing and Internet services by utilizing remote super computers time-shared over a large number of users utilizing dumb HDTV terminals. This will reduce or eliminate software costs and computer depreciation for the user at the expense of slightly poorer image quality and a user interface that takes into account the video delay associated with compression. This approach would also support the features of interactive television without the need for a local computer based terminal such as the Web-TV by Sony or set-top boxes.

Note that efficient audio and video coder-decoders (codecs) are crucial to the above scenario and we will examine the impact of the upcoming H.264 standard, a codec that promises a significant (around 50%) improvement over the existing MPEG-2 standards in terms of bit rate and video quality.

We will also examine the portable television, especially the new Pocket PC's outfitted with media player and Internet software. The Pocket PCs are gaining interest with an increasing number of Internet sites tailored to the Pocket PCs QCIF (160x120 pixel) format.

The introduction of remote computing on a decoder based terminal will be examined (Internet-computing) in light of the higher video quality of H.264 compared to its predecessors and the heavy computational load that makes hardware decoder based solutions more feasible than software ones (that require a CPU). As part of the PC-TV convergence, we present a feasibility study for developing a special user terminal called a NanoTerminal. The proposed terminal will use the latest nano-technology as CMOS is nearing its end of life as line widths scale down to the nanometer regions. Similar to current MPEG-2 based DVD players on the market, this unit would use a high-speed communications interface similar to 802.11b Wi-Fi networks but with wider range (similar to Wi-Lans Inc. WI-MAX product offerings). In addition to playing DVD movies on a DVD-ROM interface, the unit is expected to be Internet and remote computer ready.

ORGANIZATION OF THE REPORT

Internet – TV Convergence

With advances in digital hardware, portable television viewing over the Internet is possible with current commercial products such as Pocket PC and Laptop running Microsoft's Windows Media Player. We envision a convergence between Internet-TV and Internet Computing terminals where a video decoder (such as MPEG-2 and H.264) is connected to a small display (e.g. nano-Terminal). This is similar to DVD ROM players on the market today except that some type of high-speed communications link would also be present.

Pocket PC's

With the advances in digital hardware, the Pocket PC or PDA have become computationally very powerful and can run applications such as MS Office, Web Browsing and Windows Media Player. This availability of cheap computational power makes the mobile 56 Kbps connections (available with wireless GPRS services) a popular target for Pocket Television.

Motorola and Microsoft have already announced, in the Fall of 2003, a cellphone (Motorola MPx200) running Microsoft Windows principally for multimedia and scheduling applications. PDA's with 256 Mbytes removable Flash storage modules (Secure Digital Storage) can buffer a significant amount of compressed movie stream, as shown in Table. It is observed that if the playback rate is 56 kbps, 256 Mbytes of storage will enable the viewers to watch video of more than 10 hours.

Table. Secure Digital Media Capacity

Playback Rate	Playbacl time	Internet	Host-Pocket PC
(Resolution-Pixels)	(256 Mbytes store)	(Download Times)	(Download Times)
56 Kbps (160x120)	10.16 Hours	30 Minutes	3.41 Minutes
100 Kbps (240x160)	5.67 Hours	30 Minutes	3.41 Minutes
300 Kbps (320x240)	1.90 Hours	30 Minutes	3.41 Minutes

Recently introduced commercial products such as Microsofts Plus! Sync & Go which are targeted toward movies and News broadcasts over Pocket PC's could be expanded to replace direct download Internet video sites with hundreds of movies and television video in the QCIF format (160x120 pixels and above), with planned programme viewing instead of the regular scheduled formats. The "Plus Sync & Go" programme allows movies and programs to be downloaded into the Secure Digital media module when the Pocket PC's batteries are charged up each evening.

Some video cards such as the ATI All-In Wonder have supported TV and DVD viewing on personal computers with existing cable channels for a number of years but the main limitation to TV over the Internet has been bandwidth.

Although high-speed connections (e.g., ADSL) is commercially available, there are customers that are simply too far away from the central office to have a high speed line connection.

A 56 Kbps Pocket PC based GPRS connection, in addition to being extremely expensive at present, has a real time broadcasting shortcoming of momentary video freezes about once every 10 seconds, lasting about a second or two. A 56 Kbps maximum bandwidth strains the limits of conventional bit mapped compression that is achievable with single low bandwidth audio. The media format ".wmf" used by the windows media player is undergoing continual improvement, but it still suffers from the above video glitches.

Its integration with H.264 may not make much difference due to the computational loading of the Pocket PCs processor by the H.264 algorithm. There are a number of avenues open to achieving 56 Kbps Pocket Television. First, the bit mapped images can be converted into vector based images. The vector based images typically use much less bandwidth in the transmission of programme material, but the images at times are distorted. The distorted image can be re-rendered using a 3-D lighting programme such as Lightwave 7.5 but at a tremendous processing time cost, which makes it impractical. Macromedia Flash (a popular vector programme) based books that are selling this solution, overlook this problem. As Wi-Fi interface cards and hotspots are becoming increasingly available, bandwidth limitations for video media content playback will soon be a thing of the past.

LAPTOPS AND PORTABLE COMPUTERS

Today's laptop (or desktop) computers are very powerful, and higher display resolutions are available compared to the Pocket PC. However, these laptop and desktop units are easily lost or damaged. It is possible to transmit a video with 320x240 pixels screen resolution at 300 Kbps and higher bandwidths. A direct feed from a television station to the central office could be made at 1.5 Mbps ADSL rate.

A rate of about 800 Kbps is generally available from satellite service providers for on demand programming if the customer is too far from a central office ADSL/Cable connection. Note that the low bit rate MPEG-2 screen format is one quarter of the normal 6.5 Mbps bandwidth, and provides almost home theatre picture quality suitable for projection television. Wi-Fi or the wireless revolution with hotspots in many coffee shops and Internet kiosks offers IEEE 802.11b or 802.11g protocols and provides 11-56 Mbps data rates. The Wi-Fi standard is suitable for the most discriminating viewers at a screen resolution that approaches HDTV capability.

RECENT DEVELOPMENTS IN TV COMPUTING CONVERGENCE

Due to the requirement for specialized servers and softwares coupled with the general market momentum of the personal computer, there has been a general lack of interest in pursuing TV–Computer convergence. There has been some progress in the area of digital media centre applications. Here a central PC based server in the home or business would distribute music and video to the other entertainment consoles located remotely and connected via Ethernet or by Wi-Fi networks. The entertainment console would consist of a set-top box with an MPEG-2 decoder and a television set. A backchannel connection with a mouse and keyboard for media selection and Internet access would also be present.

A commercial product offering these capabilities comes from Trimedia Broadcast of the United Kingdom. The displays are connected wirelessly to a central server for media and Internet downloads. No local computers are used in this configuration with the displays, and about 20 Gbytes hard drive is used for local media storage. A similar product for the home applications comes from Prismiq tailored for music and videos.

However, it would be very useful, if a Telenet type interface similar to PC-Remote 1.1 developed by American Systems is available for remote computing applications using a mouse and clipboard editing features. The PC-Remote configuration can assume two computers; one master (remote) and one slave (local) whereas only the central server has intelligence in the proposed configuration. It is projected that due to the efforts of the IEEE's "Technical Community for Services Computing (TCSC)" project, the required servers will be available in the future and will displace the local media server in the home as the TCSC will offer more features such as remote computing with free software. The TCSC project in conjunction with a low cost Nano-Terminal should go a long way in making this computing concept a reality.

FUTURE CMOS LIMITATIONS TO NANOFABRICATION

There are major issues when considering CMOS technology for Nano fabrication of an HDTV terminal such as power management, design of new architectures, short channel limitations and fabrication costs.

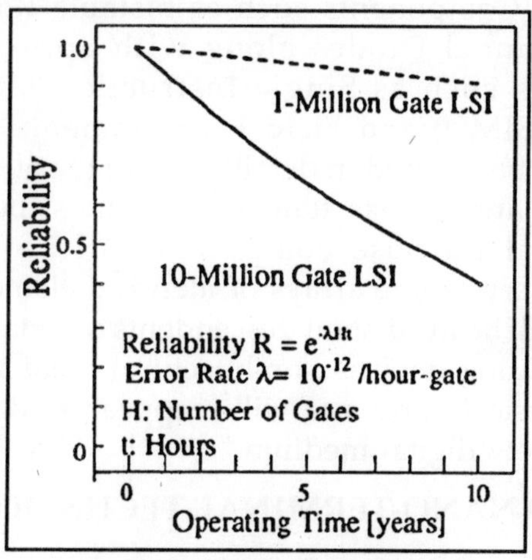

Fig. ULSI Reliability Curves

A major problem is the long term reliability of CMOS nano technology. The reliability curve developed by Shibayama et al indicates that at gate densities of ten million almost half of systems can be expected to fail within 10 years. As gate densities approach a billion (extrapolating the results predicted for 2006), we may get a 90% failure rate within 1.3 years. Error rates four orders of magnitude better than current technology are required to maintain the reliability found in a present million gate chip. The author is proposing a HDTV based nano-Terminal based on its future low cost fabrication methods such as nano-Printing on plastic substrates.

This dumb terminal would be suitable for a wide variety of Internet based media types such as movies and television as well as computing and Internet offerings. It would consist of a OLED or organic light emitting diode display (which are already in mass production) and an integrated MPEG-2 decoder logic coupled to a high speed Internet connection (56 Kbps minimum for example, through a LED interface available in upper end cellphones). We note that a single fault can make a von Neuman architecture such as Intel's Pentium 4 useless and therefore unsuitable for Nano Fabrication Technology.

In order to achieve its high density low cost structure, new memory, logic components and architectures will be considered. (Components such as bistable Rotaxanes and Resonant Tunnel Diodes along with appropriate logic architectures such as Single Instruction Multiple Data processors (SIMD) and Field Programmable Gate Arrays (FPGAs) will be covered in detail). Due to the above problems the author Margolus has stated "...our most powerful large scale general purpose computers will be built out of macroscopic crystalline arrays of identical elements.

These will be the distant descendents of today's SIMD and FPGA computing devices...architectural ideas that are used today in physical hardware will reappear as data structures within this new digital medium."

PROPOSED NANO-TERMINAL TECHNOLOGY

The proposed nano-Terminal technology uses two terminal revolutionary devices not related in any way to three terminal CMOS thin film transistors. The nanoTerminal technology has low cost manufacturing processes which are chemical and not photo-lithographic.

TWO DIMENSIONAL MOLECULAR ELECTRONIC CIRCUITS

We must consider molecular electronic circuits as rapid

progress is being made in this direction. A 2D crossbar based alternative promises simplified circuitry with fault tolerance in a Chemically Assembled Electronic Nanocomputer (CAEN).

The crossbar fabricated using nano-technology methods is important for the following reasons: A crossbar involves only two sets straight aligned wires.

Thus crossbars may be fabricated using a wide variety of techniques ranging from traditional lithography to imprinting and the chemical assembly of nano-wires.

Since a crossbar may be addressed using order (n) number of large wires to interrogate 2n nano-wires, it exhibits excellent scaling between micro and nano length scales.

The crossbar is a generic circuit that can be electronically configured for memory, logic or signal routing applications without the requirement of gain, although signal gain is eventually required.

Nano-electronic circuits that involve some level of chemical assembly in the fabrication process are unlikely to be perfect, and the crossbar structure is defect tolerant as shown by the Teramac computer.

For very high device densities, the crossbars low power consumption may outweigh other parameters such as switching speed.

A crossbar can be a naturally parallel architecture and so switching speed is not a particularly important requirement.

Fig. Rotaxane Memory Element

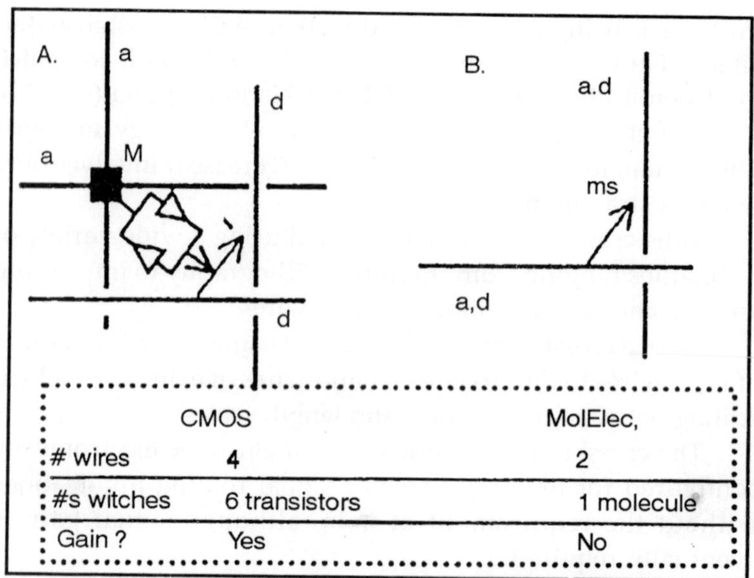

Fig. CMOS and Molecular Crossbar Switchs

In the interests of achieving point addressability and large amplitude switching within a 2D crossbar circuit Y. Luo et al describes an evolutionary progression from "bistable catenane via a bistable pseudorotaxane, to a pair of amphiphilic, bistable rotaxanes".

This scaled well from micrometer down to nanometer dimensions containing a few thousand molecules. These classes of molecules have been developed by Fraser Stoddart and his group over a 20 year period starting in 1981.

Frasers group has even shown electrochemically addressable switching and simple logic operations from some of these systems.

Equally important is that his group has demonstrated that these various classes of molecular compounds are highly modular. This means that it might be possible to have components that switch, components that aid assembly and components that aid chemical stability- all in the same molecular compound. J. Heath states. "The molecular compounds that we begin to work with, in collaboration with Stoddarts group are shown in Figure.

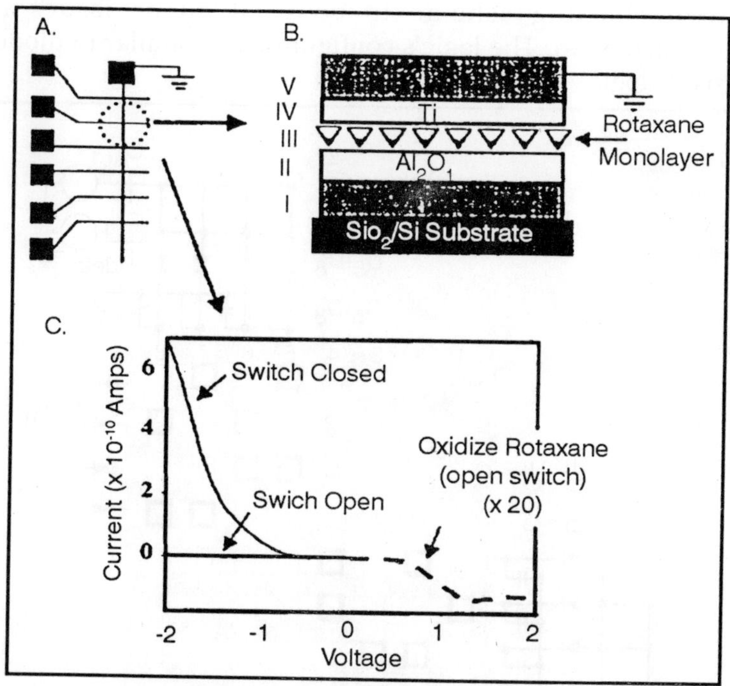

Fig. Rotaxane Layout and Switching

These molecular compounds come from a class of molecules mentioned above that are known as rotaxanes. Rotaxanes consist of two or more components mechanically interlocked with one another: one of the components is dumbbell shaped and the remaining components are rings that become trapped in the dumbbell component, encircling part of the dumbbell. In the unique class of rotaxanes used here, the dumbbell contains two bipyridinium units.

Only the two compounds of Figure that actually contain the rings are proper rotaxanes. The encircling ring(s) is (are) bis-para-phenylene-34-crown-10." A 64 bit 2D crossbar random access memory circuit has been fabricated and tested along with a circuit of two coupled, complimentary 1D logic gates to fabricate an XOR function using a lookup table. Figures shows a 4 × 4 decoder circuit and memory. Dramatic progress is being made with this technology, leaving only gain and a clocking scheme to be resolved as issues in implementation. The logic's configuration is similar to diode-resistor logic.

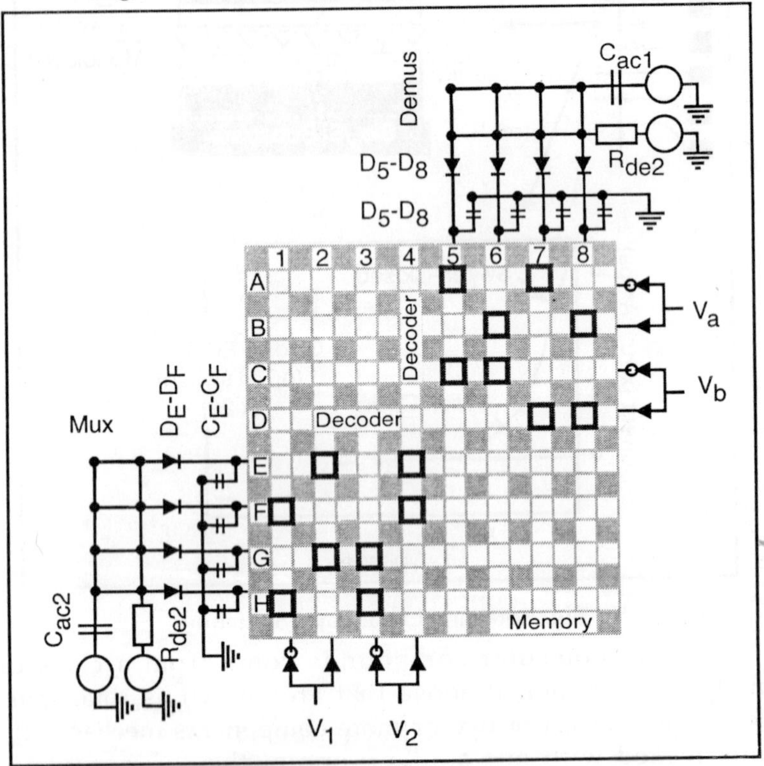

Fig. 4 × 4 Memory and Decoder

ORGANIC TRANSISTOR INTERFACING TO MOLECULAR MEMORY.

In manufacturing the logic elements needed for an MPEG-2 decoder we must consider top down (subtractive process) or bottom up (self assembly) assembly techniques whose characteristics are outlined in Table below. SIMD and especially FPGA architectures are highly regular, favoring bottom up assembly. The physical dimensions of top down fabrication is limited by its resolution ie lithography. The size limits of bottom-up self assembly could be much smaller, since assembly can be self controlled on the atomic or molecular scale. Bottom-up assembly processes will be cheaper but with higher defect rates.

Table. Assembly Characteristics

Top-Down Fabrication	*Bottom-Up Assembly*
• e.g., Lithography	• e.g., Self-Assembly
• Limited resolution	• Molecular resolution
• Expensive	• Potentially cheap
• Arbitrary structures	• Limited to regular or random structures
• Reliable 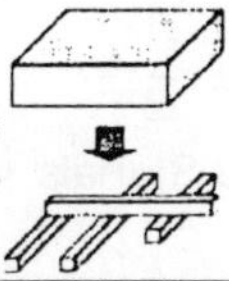	• Higher Defect Rates 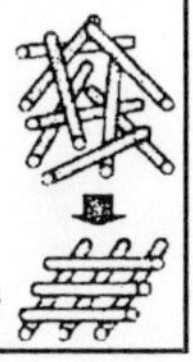

Fig. Shows the Interfacing Details of (micro) CMOS Transistors Connections which will Provide Gain and Logic Interface Functionality to a Bistable Rotaxane Molecular Memory-logic Block (nano) Connections.

The crossbar circuits were fabricated by imprint lithography. A monolayer of the rotaxane molecules was sandwiched between bottom Ti (3 nm)/Pt (5 nm) and top Ti (11 nm)/Pt (5 nm) nanowires, in fabricating a typical logic block. Figure (a) uses the junctions between the address microwires and horizontal nanowires to form a decoder. On the other hand Figure (b) uses two interface structures to create a decoder from two sets of intersecting nanowires.

While CMOS is specified, organic pentacene thin film transistors (OTFTs) could also be used with current channel lengths of 30 nm but with speed considerably slower than found with CMOS. OTFTs operate identically to CMOS except that the semiconducting layer is organic (ie Pentacene) rather than silicon. This allows a nano-Printing fabrication process rather than a top down fabrication methodology.

The circuit uses crossbar devices which sandwiches bistable molecules between wires at each junction. M. Ziegler describes the operation as follows "The result is a two terminal device that can be electrically configured to behave as a low resistance device or as a high resistance device. The devices under consideration can be switched to a low resistance state (the on state) by applying an appropriate positive voltage bias.

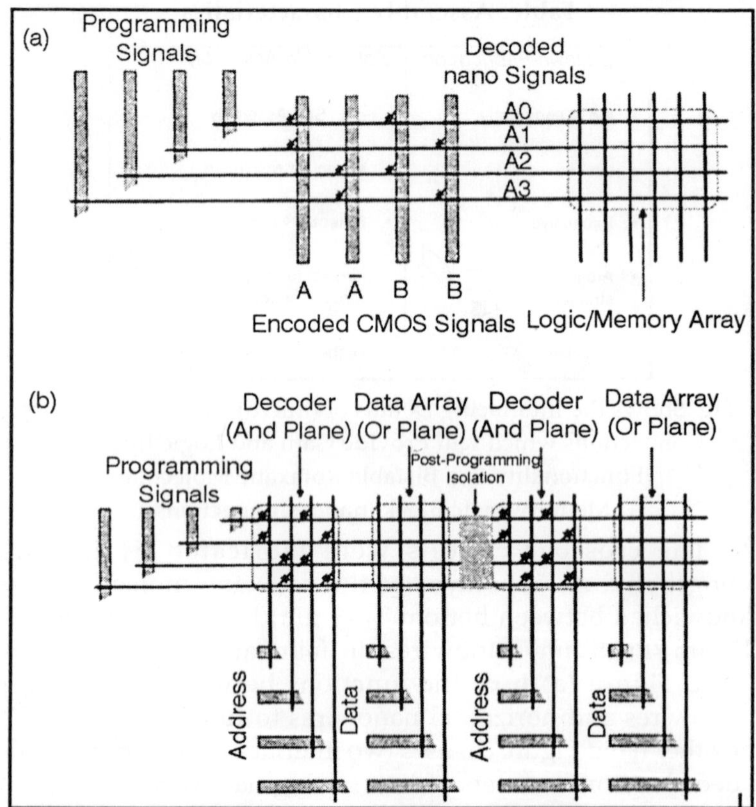

Fig. Microwire to Nanowire Interfacing

Likewise, the devices can be switched to a high resistance state (the off state) by applying an appropriate reverse voltage bias. Some of the proposed devices are one time programmable, but many can also be switched reversibly over many cycles."

Figure shows the solution to the lack of gain and inverters for molecular electronic technology. Figure is a level shifter circuit while Figure (b) is a sense amplifier, configured out of CMOS technology but applicable for OTFT technology which is operationally very similar. Figure shows a mixed CMOS/ nano full adder circuit.

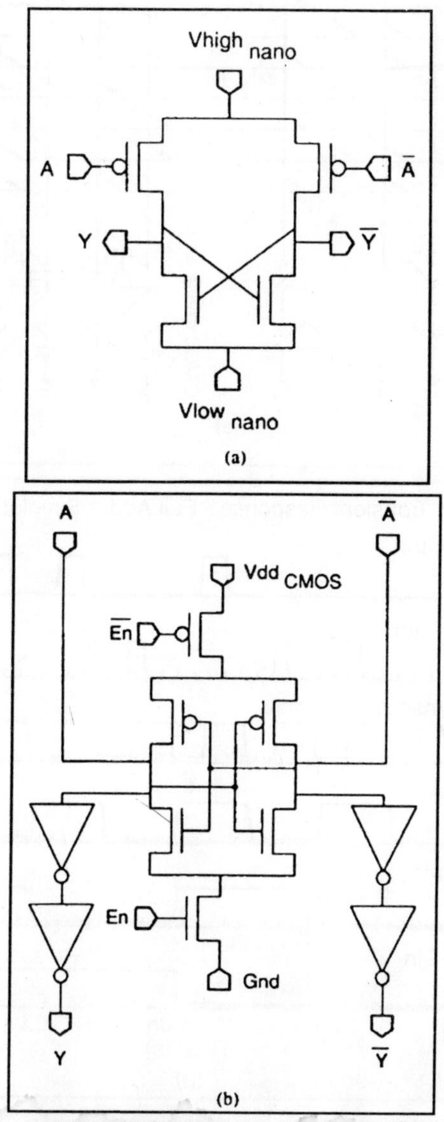

Fig. CMOS Interface Circuitry

From the performance figures tabled in M. Ziegler states

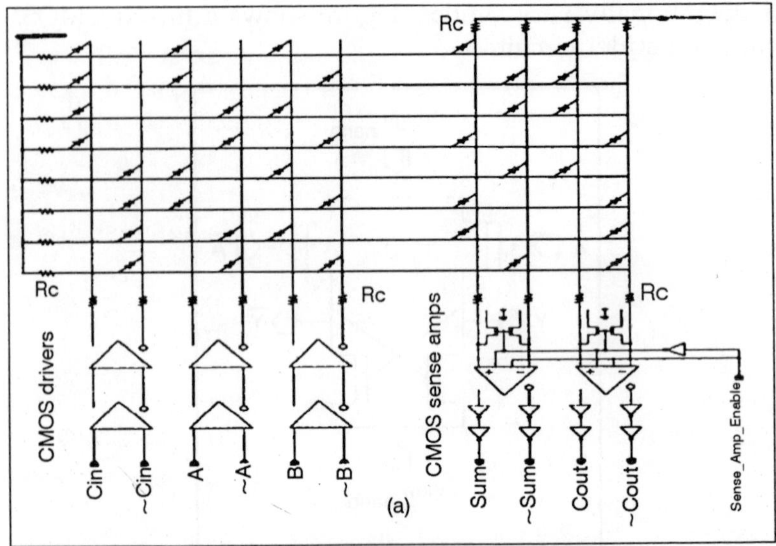

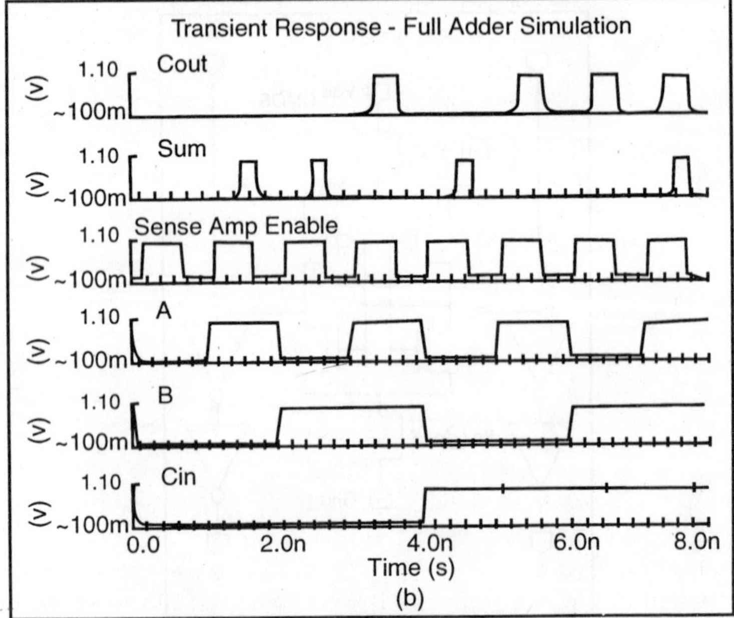

Fig. CMOS/Nano Full Adder Circuit

"it is clear that in this example, the CMOS peripheral circuitry dominates the total area, while the nano crossbar and interface structure occupy less than 4% of the total area. In comparison with a 70-nm process standard library full adder cell, we estimate the overall mixed CMOS/nano full adder to be more than twice the size of the library cell; however, considering only the nano crossbar and interface structures, the area would be only 11% of the full adder library cell.

In addition to area, the peripheral CMOS circuitry also dominated the energy consumption in this example. Only about 0.5% of the average energy is consumed by the nano crossbar, while the rest is consumed by the peripheral CMOS." As we shall see in a following section, this CMOS/molecular memory setup is a good fit for SIMD processors where large molecular memory arrays would have a small footprint.

MOLECULAR RESONANT TUNNELING DIODES (FUTURE APPLICATIONS).

Molecular Resonant Tunneling Diodes (RTDs) offer considerable layout economy compared to CMOS transistors. A single CAEN switch-diode pair will require 800 nm^2 compared to 100,000 nm^2 for a single laid-out CMOS transistor Another advantage is that if a low current CAEN signal is forced to drive a CMOS signal restoration transistor, the relatively large capacitance of the CMOS transistor will decrease the circuits overall speed compared to an RTD.

Therefore here we describe a molecular latch, constructed entirely from molecular scale devices that can perform signal restoration using power from the clock to provide gain. The molecular latch provides signal restoration, I/O isolation and noise immunity.

The molecular latch is constructed from a pair of molecular resonant tunneling diodes (RTDs). Figure shows an I-V curve of a representative RTD. The key feature of an RTD is a region of negative differential resistance (NDR), where the tunneling current falls as the voltage increases. The ratio of the current at the beginning of the NDR region to the current at the end is called the peak-to-valley ratio (PVR). Molecular RTDs that operate at room temperature with PVRs of 10 or more have already been fabricated.

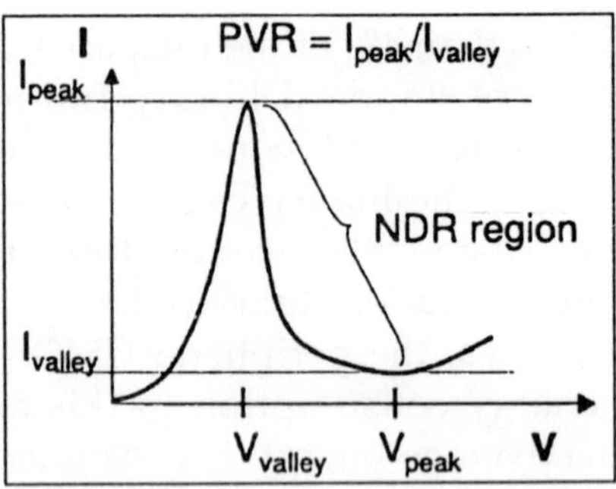

Fig. I-V Curve for RTD Device

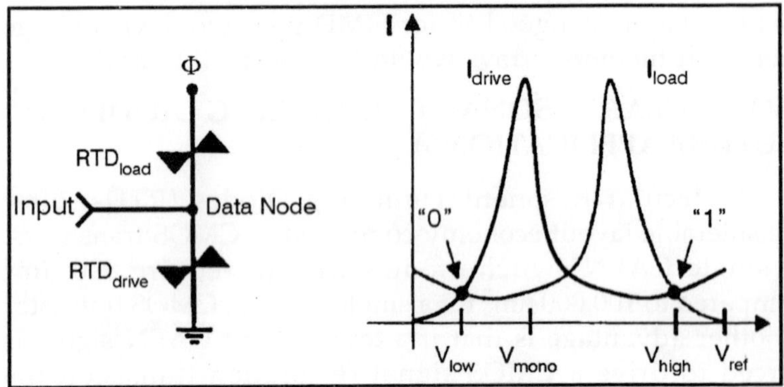

Fig. Basic Molecular Latch Configuration

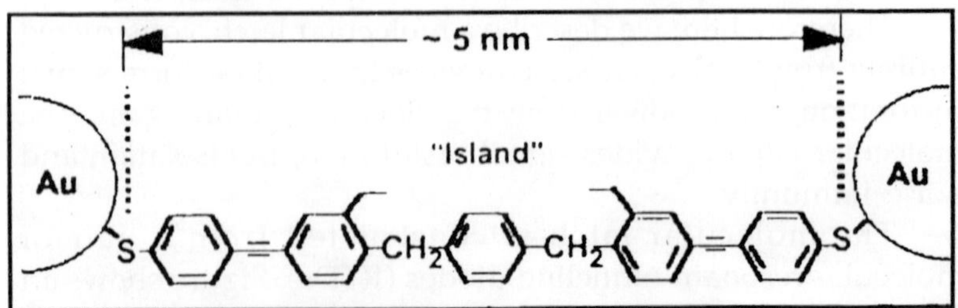

Fig. Molecular RTD Construction

Figs hows the arrangement of the RTDs in a molecular latch and a load line diagram for the molecular latch at voltage Vref. The state of the latch is determined by the voltage at the node between the two RTDs. The load line diagram shows two stable states, Vlow and Vhigh, and a third meta-stable state.

Small voltage fluctuations in the meta-stable state will push the circuit into one or the other of the stable states. The low state represents a binary "0" and the high state a binary "1".

The circuit is clocked so that the state of the latch is changed by temporarily disrupting and restoring this bistable equilibrium state. The bias voltage is temporarily lowered to Vmono, causing a shift of the load line to the left such that the circuit has only one stable state. The bias voltage is then returned to Vref and the current from the data node is able to re-establish the high-low binary states. Figure shows the construction of a molecular RTD based on the synthesis by Tour and demonstration by Reed as described in the article. "The polyphenylene-based molecular RTD of Reed and Tour uses a Tour wire backbone and is made by inserting two aliphatic methylene groups into the wire on either side of a single aliphatic ring. Since the aliphatic groups act as insulators, they become energy barriers to electron flow.

The aromatic ring between them is established as a narrow "island". The island has a lower potential energy level and electrons must pass this in order to travel the length of the wire. The molecular RTD is attached at its ends to gold electrodes by thiol (-SH) groups that absorb to a gold lattice. Tour calls molecules that bind tightly to metals "molecular alligator clips." The most common molecular-metal alligator clip is the thiol on gold arrangement." Figure shows a bistable rotaxane array with in line NDRs that can be used as a logic element or PLA type device called a nanoBlock. Figure shows these PLAs organized into an array able to perform MPEG-2 decoding using a nanoFabric sea of logic elements.

Each nanoBlock is based around a molecular logic array (MLA). At each intersection of the MLA is a reconfigurable switch (e.g. a bistable-rotaxane device) in series with a diode. Diode-resistor logic is used to perform logical operations. To create a complete logic family, signals and their complements are brought into each circuit and produce both the desired functions and their complements. Logic values are restored using molecular latchs, which also provide a mechanism for latching values and isolating outputs from one nanoBlock from the inputs of another nanoBlock. The nanoBlocks are grouped together into clusters and arranged so that the outputs of a nanoBlock intersect the inputs of two other NanoBlocks.

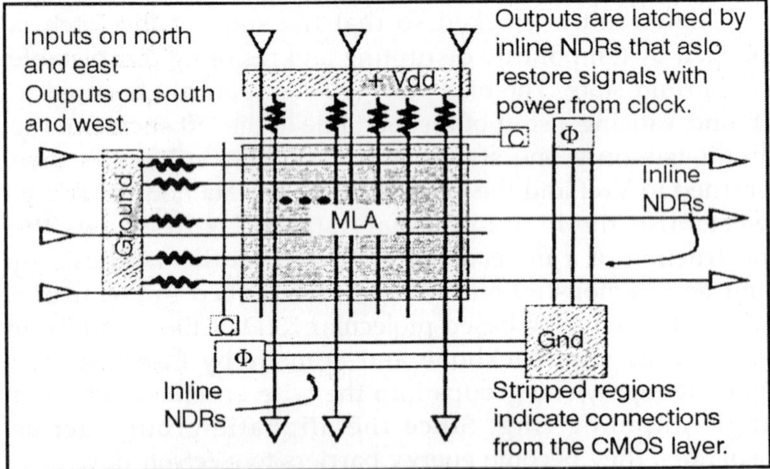

Fig. A nanoBlock Circuit Element

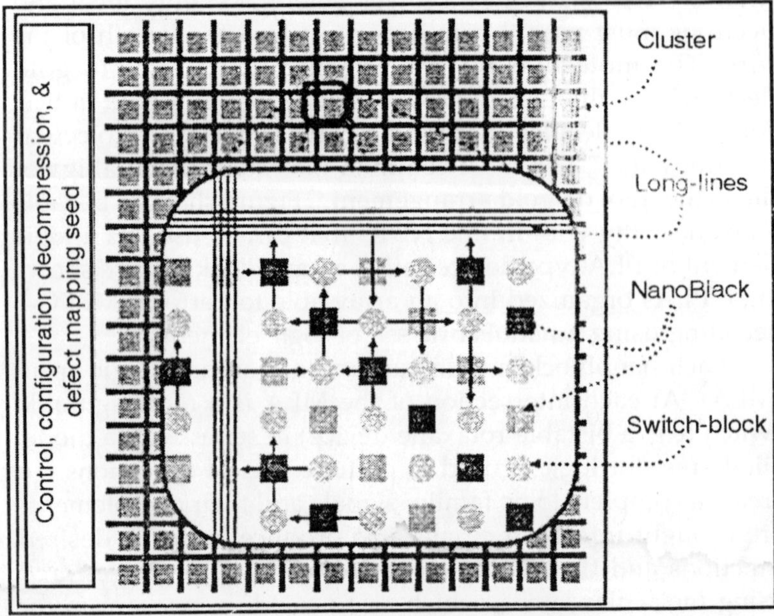

Fig. The nanoFabric of nanoBlocks

The area where the wires for four nanoBlocks intersect (two outputs, two inputs) is called a switchblock. The result is a 2-D mesh of nanoBlocks. This will fit in well with future FPGA implementations of MPEG-2 decoders where there will be a migration from micro-scale to nano-scale features. (As OTFT interface technology is replaced by molecular RTD latchs in the nanoFabric arrays of nanoBlocks).

PROPOSED TV-COMPUTER INTEGRATION

INTERNET – COMPUTING CONVERGENCE

Although, PCs are very useful in daily life, they have several limitations, such as low reliability, rapid obsolescence, high software costs and the under utilization of their full processing power. Alternatives such as the network computers are not cost effective compared to personal computers. In this section, we propose an architecture for using an HDTV monitor as a computer terminal that addresses these concerns. The proposed architecture is illustrated in Figure.

The computer raster corresponding to the remote computer server is compressed using an MPEG-2 encoder prior to being sent over a high-speed link to the HDTV set. We note that the MPEG-2 decoder shown in Figure in the user side is built-in to the HDTV (or available in a set-top box).

The television and DVD signals from the server side are fed directly into the broadband link as the bit stream is already in MPEG-2 format. The mouse and keyboard would be the only added accessories required at the users side along with the HDTV.

The signals from the keyboard and mouse travel via a back channel from the broadband link to the parallel computer as shown in Figure. The traffic from the remote server to the home terminal (i.e., the front channel) would be much higher compared to the back channel. Hence, an asymmetrical link with front channel capacity of a few Mbps (although 6.5 Mbps is typically specified for a DVD quality pay TV channel, this is being rapidly reduced to under 1.0 Mbps with H.264) and a back channel capacity of a few tens of Kbps would be sufficient.

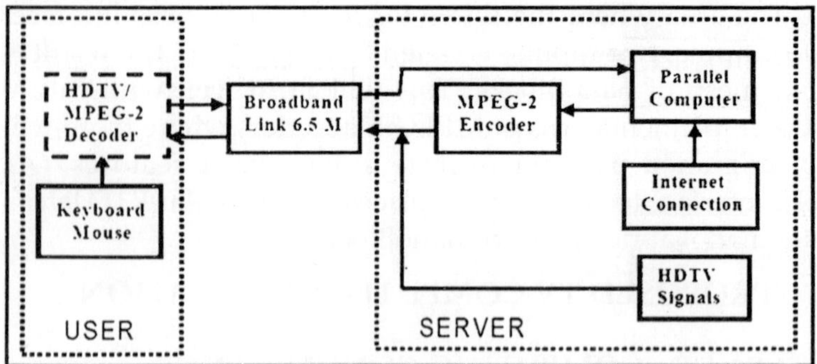

Fig. The Proposed Architecture

HDTV REMOTE TERMINAL

In order to make the HDTV monitor suitable for use as a computer terminal, we have to ensure an acceptable quality for the text character resolution. We note that there is a limit to the minimum character size (i.e., font-size in points), resolvable with the addition of artifacts introduced by video compression. There are two alternatives to address the resolution concerns.

In the first alternative, we can limit the character size to a minimum font size. It has been found experimentally that a font size of at least 8 would provide text with acceptable quality. In the second alternative, the character stream can be sent separately with a local decoder similar to the closed caption available in the present television system.

In addition to an acceptable subjective quality of the text characters, the video delay between the remote server and the local TV should be minimized. The delay occurs mainly due to the time required for video compression, transmission over networks, and decompression. Typical video delay would rule out most interactive game applications, which rely on immediate response from the users. However this is not a serious limitation since dedicated game consoles are generally used for such applications.

Office applications and the Internet can be supported with some hardware modifications to allow for direct mouse cursor movements on the screen. The proposed architecture requires no modification to the applications software (such as word processors and spreadsheets) used at the server. However minor modifications in the server operating system (such as Linux, allowing the remote mouse and keyboard events to be processed in a read-write event queue) would be required. The MPEG-2 encoder and decoder software was modified first to enable user data to be displayed on the HDTV monitor.

This was further modified to send mouse and cursor positions to the server. Sending the current terminal window data by a mouse click on the image would also send an updated cursor position right after the text data. This would require the need for a special cursor position frame header to indicate button click events. It is expected that the user data will be sent after the encoding and transmission of the MPEG-2 compressed video screen.

Fig. Compression of the Raster Image Using MPEG-2 Codec

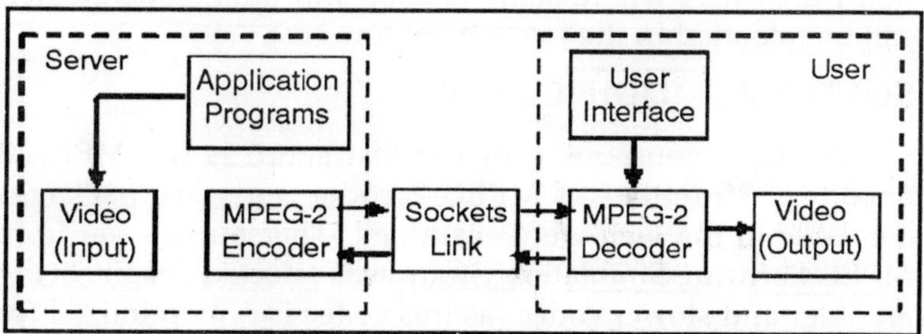

Fig. Modified MPEG-2 Software Schematic

HARDWARE MODIFICATIONS

The HDTV monitor would require a small peripheral circuit to allow its built in microprocessor controller or FPGA decoder to handle raster-op (BitBlt- a state machine based video processor performing logic operations between a source bitmap and video memory) operations for local mouse movement and keyboard character display (buffered prior to being sent back to the server) as shown in Figure. This is why the mouse and keyboard cannot be connected to a set top box unless it contains the decoder functions.

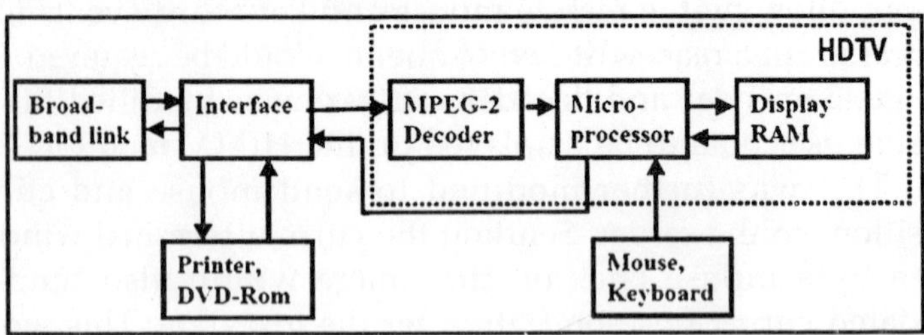

Fig. Peripheral Interfaces to HDTV There would be no high level intelligence at the HDTV based terminal, rather massively parallel supercomputers would serve a large number of users with time shared MPEG-2 encoding of different rasters.

This architecture would allow remote supercomputer power to be available to even Personal Digital Assistants and network appliances via a wireless connection protocols such as GPRS service. This application is especially suitable for the non-PC owners where some latency and image degradation can be tolerated in the interests of low cost.

SOFTWARE MODIFICATIONS

A major component of the test setup is the MPEG-2 encoder. A number of MPEG-2 codec software packages available in the web were examined. The source code from MPEG Software Simulation Group was selected despite being an older and slower codec (as this codec is not optimized for Intel MMX instructions, etc). It is a highly modular encoder-decoder package despite having some major shortcomings in the X11 display (e.g., Linux 24 bit display terminals would not work with this package).

The codec was easily modified to achieve the desired objectives of a testbed to develop software for the proposed architecture. In the following sections we outline approaches to minimize the lossy nature of MPEG-2 compression and decompression as well as minimizing the problems associated with the video delay. We note that the cursor positions of the local monitor have to be sent to the remote computer periodically. To allow this, the MPEG-2 encoder and decoder are connected via a TCP/IP Ethernet socket client. As per the listing, a flag "ETHERNET" must be defined in each of these programs to replace the file I/O with Ethernet based direct connections. The modifications are sophisticated enough so that a URL numerical code is replaced by client/server computer names.

These are put in the parameter list of the commands replacing the file names to allow maximum versatility. The MPEG-2 decoder display programme was re-written to achieve 24 bit colour display functionality on Linux machines (which was on the MPEG-2 softwares to-do list). The MPEG-2 Software Simulation Group's recommendation to use the Berkeley MPEG decoder software interface was explored along with a software package for X11 from Karel Tsuda from Japan.

It was found that Gnome/GTK graphics programming provides a more versatile and user-friendly interface. This resulted in the addition of a "GNOME" editing screen for the user data to be edited prior to being sent back to the encoder programme. This was done by a left button mouse click on the image screen. The left button is also used by this handler for cut and paste operations (on the data) in the editing pane integrated with the image screen. A mouse click on the display screen also sends back to the server the mouse cursor co-ordinates along with the type of mouse button click.

H.264 CODEC IMPLICATIONS

In the previous section, we presented a schematic for integrating HDTV with remote computing servers. This schematic was primarily based on the MPEG-2 coding standard, which has been adopted for HDTV broadcasting and DVD movies. However, recently a new coding standard is being developed which is expected to provide a significant performance enhancement over the MPEG-2 standard. The H.264 offers MPEG-2 quality video at under 1 Mbps bandwidth allowing a minimum 2.0 GHz Pentium 4 with 256 Mbytes of storage to serve as an H.264 decoder and file manager over a high speed ADSL Internet connection. A projector with this setup could offer wall sized screen home theatre coupled with a high quality Lucasfilm THX 7.1 audio system.

Although, H.264 standard provides a superior performance, it is not likely to be incorporated in the digital TV in the near future. However, we can still eliminate the PC upgrade requirements completely by combining storage and computing functions at a central server through H.264 (or MPEG-2) coded Internet pipe with a set-top box decoder and video projector at the consumer end. For home use, it turns out that it is possible that regular television is feasible for H.264 compression based on its higher video quality when compressing computer screens containing Word or Excel documents.

The idea is applicable for pocket PC's whose H.264 demo video clips show a noticeable shortfall in processing power for medium range pocket PC's (Intel 300 MHz ARM). Here H.264s complexity in achieving a typical 50% bandwidth reduction when compared to MPEG-2 taxes its processor to the point that just an H.264 hardware decoder is proposed for a terminal. While MPEG-2 has the present HDTV and commercial markets wrapped up, there is room for H.264 in the Internet Television Broadcasting area where Microsofts.wmf (similar to MPEG-4) has the market.

Microsoft is expected to migrate to H.264 as they have contributed significantly in the development of this format. This is particularly welcomed by bandwidth-limited consumers (56Kbps dial-up, GPRS etc) who wish to obtain a complete multi-media experience in PC-TV integration. The proposed PC-TV integration concept, in addition to avoiding a processor upgrade, would solve the middle-ware problems presently plaguing the industry. This eliminates the concerns about copyright violations as all buffered media content would be under the copyright owners possible supervision.

PERFORMANCE EVALUATION

In this section, the performance of the proposed

architecture is evaluated. The HDTV standard typically allows video with two resolutions, a progressive scanned 1080 x 720 pixels (width x height), and an interlaced 1920 x 1080 pixels. Due to the unavailability of an actual HDTV, we have simulated the experiments with a Linux workstation. For testing purposes only 640 x 480 pixel images are used since this is the worst case as far as resolution of de-compressed images is concerned and furthermore, the software codec used can only support this resolution at this time.

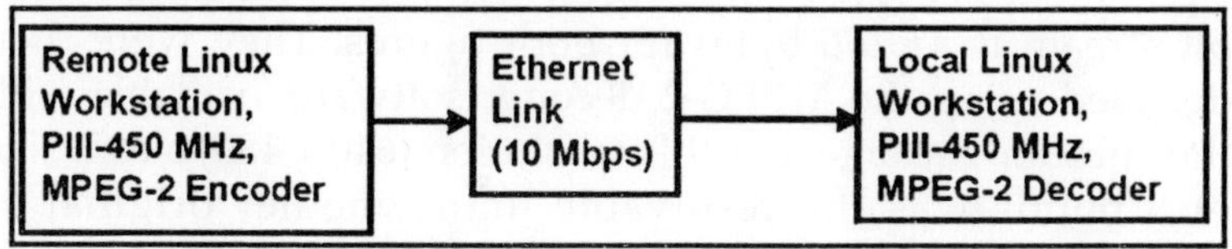

Fig. Experimental Setup.

Here we present two approaches for minimizing the compression artifacts. The first approach is to encode the entire 640 x 480 pixels screen with a mix of characters and graphics. Typical screens of a Microsoft Word document at various character point sizes were first compressed using the MPEG-2 encoding software mentioned above.

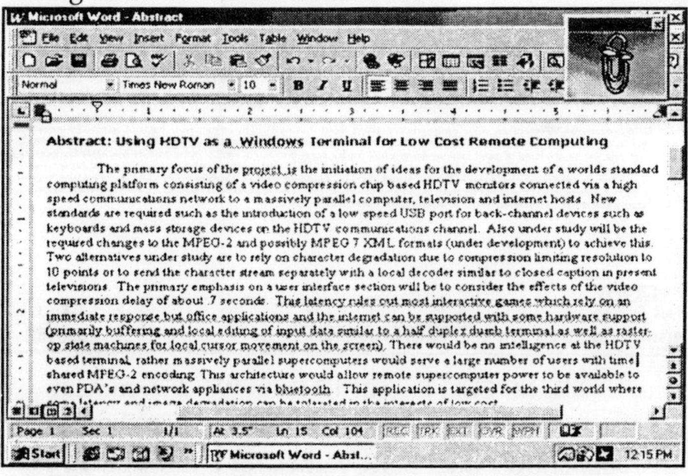

Abstract: Using HDTV as a Windows Terminal for Low Cost Remote Computing

The primary focus of the project is the initiation of ideas for the development of a worlds standard computing platform consisting of a video compression chip based HDTV monitors connected via a high speed communications network to a massively parallel computer, television and internet hosts. New standards are required such as the introduction of a low speed USB port for back-channel devices such as keyboards and mass storage devices on the HDTV communications channel. Also under study will be the required changes to the MPEG-2 and possibly MPEG 7 XML formats (under development) to achieve this. Two alternatives under study are to rely on character degradation due to compression limiting resolution to 10 points or to send the character stream separately with a local decoder similar to closed caption in present televisions. The primary emphasis on a user interface section will be to consider the effects of the video compression delay of about .7 seconds. This latency rules out most interactive games which rely on an immediate response but office applications and the internet can be supported with some hardware support (primarily buffering and local editing of input data similar to a half duplex dumb terminal as well as raster-op state machines for local cursor movement on the screen). There would be no intelligence at the HDTV based terminal, rather massively parallel supercomputers would serve a large number of users with time shared MPEG-2 encoding. This architecture would allow remote supercomputer power to be available to even PDA's and network appliances via bluetooth. This application is targeted for the third world where

Fig. MPEG-2 Encoded 10 Point Word Document

A windows programme 'CAPEX' was used to compress the windows image from a 640 x 480 pixel 256 colour screen into a 'GIF' bitmap, which was then transferred to a Linux desktop PC. A windows programme was used since the Linux equivalent 'Magicapture' is still under development with compile bugs preventing its use. In the Linux environment, the 'XV' programme is used to convert the 'GIF' file to 'PPM' format, which is required by the MPEG-2 encoder.

Several Microsoft 'Word' images were encoded. Figure shows an image with 10- point font size. A similar image with 8 point font size is shown in Figure. The size of the compressed bit stream is 41,266 bytes for both figures. They were then decoded using the MPEG-2 decoder software available with the encoder package. For this resolution (640 x 480) a font size of 8 points was the resolvable limit whether original or reconstructed. It is observed that the reconstructed image has artifacts in the form of a gray cloud around the text paragraphs, which does not degrade the readability of the text.

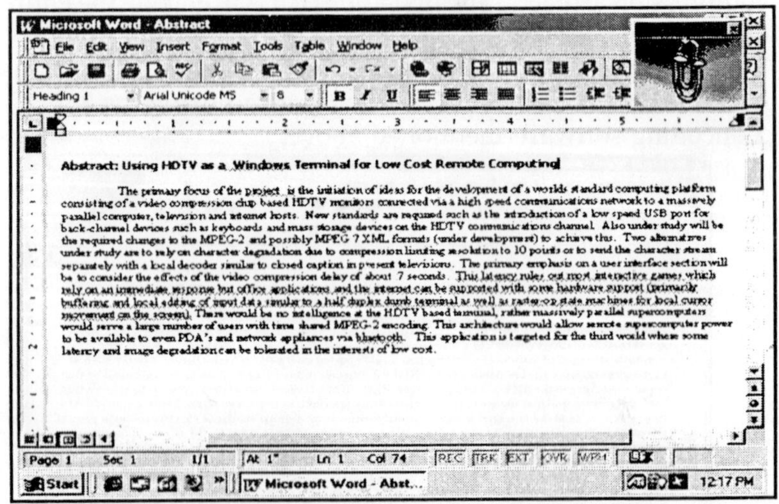

Abstract: Using HDTV as a Windows Terminal for Low Cost Remote Computing

The primary focus of the project is the initiation of ideas for the development of a worlds standard computing platform consisting of a video compression chip based HDTV monitors connected via a high speed communications network to a massively parallel computer, television and internet hosts. New standards are required such as the introduction of a low speed USB port for back-channel devices such as keyboards and mass storage devices on the HDTV communications channel. Also under study will be the required changes to the MPEG-2 and possibly MPEG 7 XML formats (under development) to achieve this. Two alternatives under study are to rely on character degradation due to compression limiting resolution to 10 points or to send the character stream separately with a local decoder simular to closed caption in present televisions. The primary emphasis on a user interface section will be to consider the effects of the video compression delay of about 7 seconds. This latency rules out most interactive games which rely on an immediate response but office applications and the internet can be supported with some hardware support (primarily buffering and local editing of input data simular to a half duplex dumb terminal as well as raster-op state machines for local cursor movement on the screen). There would be no intelligence at the HDTV based terminal, rather massively parallel supercomputers would serve a large number of users with time shared MPEG-2 encoding. This architecture would allow remote supercomputer power to be available to even PDA's and network appliances via bluetooth. This application is targeted for the third would where some latency and image degradation can be tolerated in the interests of low cost.

Fig. MPEG-2 Encoded 8 Point Word Document

Table PSNR Values for MPEG-2 Encoding

Table shows the PSNR values for 8 through 12 point text encoded sequentially with I-B-P frames and individually using I frames only. The IBP Frames values are worst case being sequentially encoded and decoded as a group. Blocking artifacts make odd point text values slightly lower than even point ones due to a poor block fit over the character line.

Text Point Size	I-Frame only MPEG-2 Encoding PSNR (in db)	I-B-P Frame MPEG-2 Encoding PSNT (in db)
8 Point	41.604 (acceptable)	33.621 (unacceptable)
9 Point	41.383 (acceptable)	33.598 (unacceptable)
10 Point	44.324 (good)	36.534 (marginal)
11 Point	42.012 (acceptable)	37.527 (unacceptable)
12 Point	44.240 (good)	37.708 (unacceptable)

As text in a typical document is likely to be 10-points and above, the performance of this architecture is acceptable for text transmission. This is the preferred choice due to the ease of implementation as no software applications needed to be modified. We have also evaluated the performance of H.264 codec for coding spreadsheets and word images.

Table compares the performance of H.264, and MPEG-2 codecs. It is observed that the H.264 Standard provides a superior performance. Figure shows an H.264 decoded image, which has a higher subjective quality compared to the images shown in Figure. Table PSNR (in dB) performance of H.264 and MPEG-2 for "MS Word" image frames. The bit-rate is 1.5 Mbps, and the frame rate is 15 frames/sec.

Text Point Size	H.264	MPEG-2
8 Point	51.02	41.60
9 Point	51.17	41.38
10 Point	50.71	44.32
11 Point	50.46	42.01
12 Point	50.55	44.24

Although, the words can be encoded as picture, these can also be encoded by other methods that may be much simpler. For example, the character encoding channel in the MPEG-2 specifications allows compression free characters to be locally generated. This feature can be incorporated in the proposed architecture by means of a set of character ROMs (Read only memories) in the HDTV decoder similar to closed caption text present on existing televisions. As closed caption text is likely to be present with HDTV, this would require no modifications to the decoder.

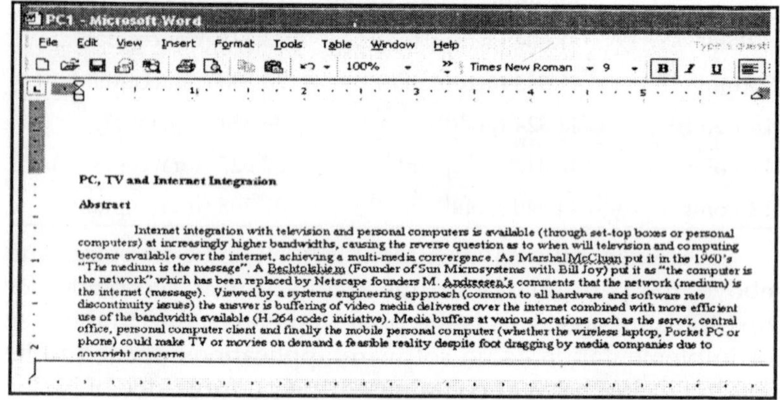

PC, TV and Internet Integration

Abstract

Internet integration with television and personal computers is available (through set-top boxes or personal computers) at increasingly higher bandwidths, causing the reverse question as to when will television and computing become available over the internet, achieving a multi-media convergence. As Marshal McCluan put it in the 1960's "The medium is the message". A Bechtolshiem (Founder of Sun Microsystems with Bill Joy) put it as "the computer is the network" which has been replaced by Netscape founders M. Andreesen's comments that the network (medium) is the internet (message). Viewed by a systems engineering approach (common to all hardware and software rate discontinuity issues) the answer is buffering of video media delivered over the internet combined with more efficient use of the bandwidth available (H.264 codec initiative). Media buffers at various locations such as the server, central office, personal computer client and finally the mobile personal computer (whether the wireless laptop, Pocket PC or phone) could make TV or movies on demand a feasible reality despite foot dragging by media companies due to copyright concerns

Fig. H.264 Encoded 9 Point Word Document

A typical computer session would have the window frame displayed using MPEG-2 de-compression and the characters would be sent over an interleaved section of the communications channel and would be displayed at a point set by the server controlled cursor.

Character updates would occur at a point set by this cursor. This will provide a much sharper image for characters to allow fine print to be read from a distance but would restrict the number of available character fonts unless a downloadable font in temporary local storage is available from the server.

Due to the need for extensive server software modifications to make this work, the added clarity was seen as not being worth the effort at this stage of development.

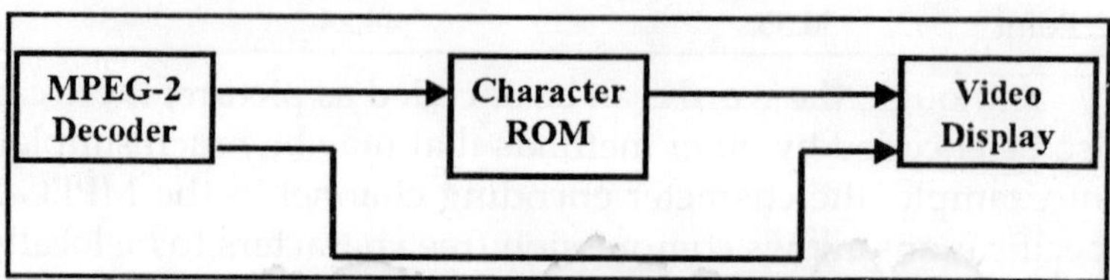

Fig. Closed Caption Text Schematic

VIDEO DELAY

In practice, when a user is working on an HDTV monitor, a delay can be expected whenever changes are made on the screen. This slow response is due to the encoding and decoding of the video frames. If this delay occurs at every character level, it will be very annoying to work with an HDTV monitor. In order to achieve a faster response, we use a buffer where the intermediate characters are stored. This would be done in a small editing window underneath the decompressed display screen. A keyboard return or a mouse click can be used to send the local buffer back to the server which will then send the next sentence under the cursor into the editing window for local editing and further input.

The line being edited will be highlighted on the image screen, and the remote editor will function as a line editor as one line is processed at a time. The small local terminal window at the bottom of the screen used for editing would also respond to selected control characters used to enable a "Terminal" mode to be used for applications such as "calling up files from an FTP site", and "performing searches". Commands (line input) from the terminal mode can be sent to the central host which would be aware of the special mode that the terminal window is in.

The treatment of an FTP file will depend on the files suffix i.e. the *graphics* file would be sent via MPEG-2 compressed video to a small local window where they could be resized and entered into the document by being mouse selected and placed through a simple graphics editor. Text files would respond to cut and paste commands ^c and ^v after the appropriate text has been selected by the mouse to allow insertion into the document being processed. It is felt that this is a more versatile and flexible setup than found with current windows based editors. The mouse movements would be handled by a local raster-op state machine to give an immediate response for a new cursor location.

Only mouse clicks giving a current windows position would be sent back to the remote server for processing and updating. The remote server would take the mouse positions with click type codes and generate an event to the server software requiring operating system modifications. Except for mouse display and control, the HDTV would act like a dumb windows terminal with all processing for word processing, spreadsheet and other windows (text box) based character entry handled by the remote server with little or no software modification to current applications.

The MPEG-2 encoder in the remote servers would be active only after updating the screen after a mouse click or keyboard cursor control function, and an MPEG-2 encoder could handle a large number of users on a time shared basis. The quality of the reconstructed rasters depends significantly on the coding strategy. The MPEG-2 standard uses a group of picture (GOP) concept, and encodes intra-coded (I), predicted (P), and bi-directionally predicted (B) frames. Although, the P and B frames increases the compression performance, it also increases the complexity and the delay.

For most applications, encoding all frames as I frames will simplify the coder design. Le et al. showed that more than 70 I frames can be encoded (using JPEG) per second using a simple C*RAM based encoder. As we will show below 'I-frame only' encoding will result in a maximum data rate of 256 Kbps per user and will greatly simplify the encoding of different users images as previous images need not be saved. The software used to evaluate the architecture is available at a web site for download. The images used and the modified configuration files for the MPEG-2 encoder are also included.

Parallel Remote Server Configuration

The remote servers are expected to have a large number of users accessing the servers simultaneously. Therefore, the servers need to have a large amount of processing power. A LINUX based parallel server with an array of high-speed processors arranged in a N by N node configuration is a potential choice. This server can also exploit the existing LINUX clustering software. Each processing node would have an MPEG-2 encoder card for compressing the users windows session and a four channel 1 Gbit Ethernet card for N by N node connections, mass storage and fault tolerance. The MPEG-2 card should support 76 users minimum (assuming an update each second/user) and for a 640 by 480 colour image compressed by a factor of 30 (JPEG I frames) would require an I/O bandwidth of 256 Kbps. Assuming a 10 by 10 by 10 array of 1024 processors in a system with the maximum bandwidth of each node to be 1 Gbps/ 10 = 100 Mbps, each node should be able to support 200 users per node assuming half of the bandwidth is for disk I/O. The hard drive storage would be in a RAID format as per existing (Hitachi-HP "Freedom Storage" products) connected to the processing nodes through SNMP protocol Ethernet channels as per existing network servers.

There would be several processors in a rack configured by a built in router to replace a defective node in the local N by N configuration (for fault tolerance). As the 256K/user bandwidth is much less than MPEG-2 requirements for full motion video (6.5 Mbps), the computing section could be multiplexed or even separated from the video channel. This would offer a dual channel capability that could result in an ADSL lite mode for the computer services with longer distances to the peg at the telephone companies' central office (ADSL is 1.5-8 Mbps at present). The server would be allocated to business users during the day, home users during the early evening hours with scientific super computing projects taking place during the late evening and early morning times.

In this chapter, we have presented an architecture for PC – TV convergence. We showed that there are a few minor inconveniences, but these can be eliminated through user interface adaptations. We presented the performance evaluation of several known coding standards, such as MPEG-2 and H.264. The subjective quality of the decoded images is acceptable, and the proposed architecture looks feasible.

HDTV NANO-TERMINAL: FEASIBILITY STUDY

In the previous chapter, we presented a schematic of TV-Computer integration. With the advances in computing and nano technology, the shape of computers and TVs is likely to change in future. In this chapter, we investigate potential technologies that will of great interest in the PC-TV integration. As shown in Figure we will consider the details in implementing a user terminal using nano-technology.

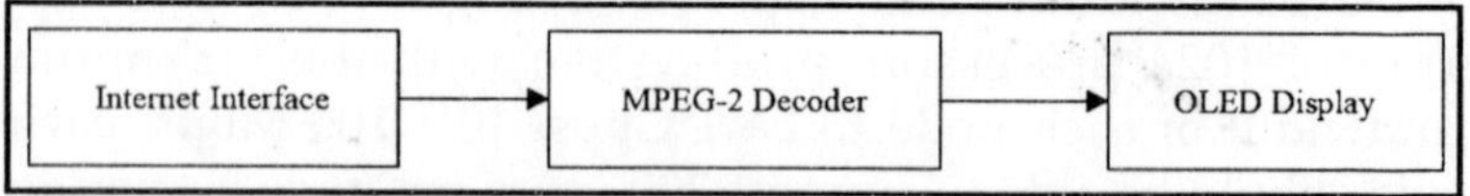

Fig. Nano-Terminal Block Diagram

The organization of the chapter is as follows.

APPLICATIONS: MPEG-2 ENCODING SOFTWARE

We have shown the molecular transistor building blocks required for our application. Now we will consider the complexity of the MPEG-2 algorithm that can be used. A schematic of the MPEG-2 algorithm is shown in Figure. The MPEG-2 algorithm consists of the 2-D 8x8 pixel block forward discrete cosine transform (FDCT) of an error frame, followed by a quantization (Q) and resultant variable length coding (VLC) of the error signal.

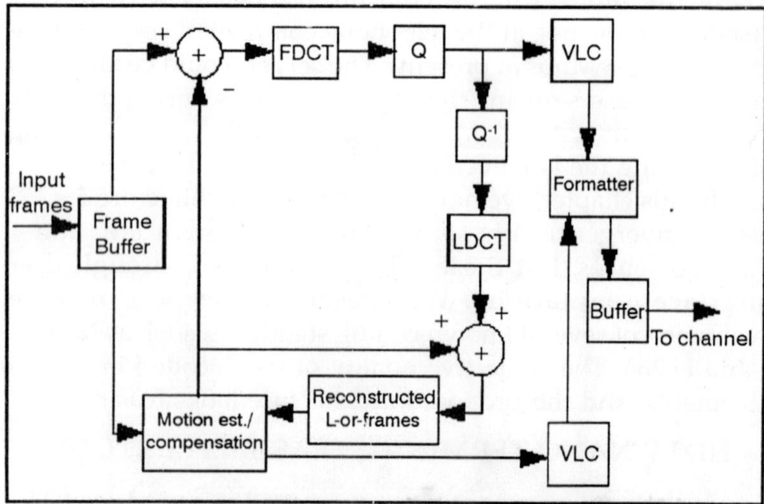

Fig. MPEG-2 Encoder Block Diagram

The inverse quantization and inverse discrete cosine

universities and national research institutes, as well as via the Japan Society for Promotion of Science (JSPS). The most active programs are those at Tokyo University, Kyoto University, Tokyo Institute of Technology, Tohoku University, and Osaka University. The Institute of Molecular Science and the Exploratory Research on Novel Artificial Materials and Substances programme promote new research ideas for next-generation industries (5-year university-industry research projects).

The "Research for the Future" initiative sponsored by JSPS has a programme on "nanostructurally controlled spin-dependent quantum effects" at Tohoku University. Monbusho's funding contribution to nanotechnology programs is estimated at ~ $25 million. In total, MITI, STA, and Monbusho allocated ~ $120 million for nanotechnology in 1996.

Large companies also drive nanotechnology research in Japan. Important research efforts are at six institutions: Hitachi (Central R&D Laboratories, where nanotechnology is ~ 25% of long-term research)—Hitachi has ~ $70 billion per year in sales; NEC (Fundamental Research Laboratories, where nanotechnology is estimated to be ~ 50% of the precompetitive research)—NEC has ~ $40 billion per year in sales; NTT (Atsugi Lab); Fujitsu (Quantum Electron Devices Lab); Sony; and Fuji Photo Film Co. An allocation of 10% of sales for research and development is customary in these companies, with ~ 10% of this for long-term research.

Some Japanese nanostructured products already have considerable market impact. Nihon Shinku Gijutsu (ULVAC) produces over $4 million per year in sales of particles for electronics, optics, and arts. Also, there are in Japan, as in the United States, private consortia making an increased contribution to nanotechnology R&D:

- The Semiconductor Industry Research Institute of Japan (SIRI), established in 1994, focuses on long-term research with partial government funding
- Semiconductor Leading Edge Technologies, Inc. (SELETE), established by ten large Japanese

semiconductor companies in 1996, focuses on applied research and development with an estimated budget of $60 million in 1997

- Semiconductor Technology Academic Research Centre (STARC) promotes industry-university interactions

Strengthening of the nanotechnology research infrastructure in the last years has been fueled by both the overall increase of government funding for basic research and by larger numbers of academic and industry researchers choosing nanotechnology as their primary field of research. Potential industrial applications provide a strong stimulus. A systems approach has been adopted in most laboratory projects, including multiple characterization methods and processing techniques.

A special Japanese research strength is instrumentation development. The university-industry interaction is stimulated by the new MITI projects awarded to universities in the last few years that encourage temporary hiring of research personnel from industry. Other issues currently being addressed are more extensive use of peer review, promoting personnel mobility and intellectual independence, rewarding researchers for patents, promoting interdisciplinary and international interactions, and better use of the physical infrastructure.

China

Nanoscience and nanotechnology have received increased attention in China since the mid-1980s. Approximately 3,000 researchers there now contribute to this field. The ten-year "Climbing Project on Nanometer Science" (1990-1999) and a series of advanced materials research projects are the core activities.

The Chinese Academy of Sciences sponsors relatively large groups, while the China National Science Foundation (CNSF) provides support mainly for individual research projects. Areas of strength are development of nanoprobes and manufacturing processes using nanotubes. The Chinese

Physics Society and the Chinese Society of Particuology are societies involved in the dissemination of nanotechnology research.

India

India's main research activities are on nanostructured materials and electronic devices. These involve a combination of research institutions (the Central Electronics Engineering Research Institute in Pilani, the Space Application Centre in Ahmedabad, and others), funding organizations (the Centre for Development of Materials in Pune, the Indian Institute of Science in Bangalore, and others), and industry.

Taiwan

Taiwan's major nanotechnology research effort is conducted in the area of miniaturization of electronic circuits. The research is conducted in academic institutions and at the Industrial Technology Research Institute. Government funds for fundamental research are channeled via the National Science Council.

South Korea

A national research focus on nanotechnology was established in Korea in 1995. The Electronics and Telecommunications Research Institute (ETRI) in Taejon, Korea's science city, targets advanced technologies for information and computer infrastructures, with a focus on nanotechnology. The emphasis is on nanoscale semiconductor devices and particularly on semiconductor quantum nanostructures and device applications (lasers, modulators, switches and logical devices, resonant tunneling devices, self-assembled nanosize dots, single-electron transistors, and quantum wires).

Singapore

Nanotechnology research received a considerable boost in Singapore by the initiation of a national programme in this area in 1995.

Australia

The National Research Council (NRC) of Australia has sponsored R&D in nanotechnology since 1993. Research groups work on synthesis of nanoparticles for membranes and catalysts (University of New South Wales), nanofiltration (UNESCO Centre for Membrane Science and Technology), and use of nanoparticles in processing minerals for special products (the Advanced Mineral Products Research Centre at the University of Melbourne). AWA Electronics in Homebush has the largest industrial research facility for nanoelectronics in Australia.

EUROPE

European Community (EC)

The term "nanotechnology" is frequently defined in Europe as "the direct control of atoms and molecules" for materials and devices. A more specific definition from H. Rohrer is a "one-to-one relationship between a nano-object or nano-part of an object and another nano-, micro- or bulk object." The nanotechnology field as defined in this WTEC report includes these aspects, with the clarification that only the specific, distinctive properties and phenomena manifesting at length scales between individual atoms/molecules and bulk behaviour are considered.

There are a combination of national programs, collaborative European (mostly EC) networks, and large corporations that fund nanotechnology research in Europe. Multinational European programs include the following:

- The ESPRIT Advanced Research Initiative in Microelectronics and the BRITE/EURAM projects on materials science in the EC are partially dedicated to nanotechnology.
- The PHANTOMS (Physics and Technology of Mesoscale Systems) programme is a network created in 1992 with about 40 members to stimulate nanoelectronics, nanofabrication, optoelectronics, and electronic switching. Its coordinating centre is at the

IMEC Centre for Microelectronics in Leuven, Belgium.

- The European Science Foundation has sponsored a network since 1995 for Vapour-phase Synthesis and Processing of Nanoparticle Materials (NANO) in order to promote bridges between the aerosol and materials science communities working on nanoparticles. The NANO network includes 18 research centers and is codirected by Duisburg University and Delft University of Technology.
- The European Consortium on NanoMaterials (ECNM) was formed in 1996, with its coordinating centre in Lausanne, Switzerland. This group aims at fundamental research to solve technological problems for nanomaterials and at improved communication between researchers and industry.
- NEOME (Network for Excellence on Organic Materials for Electronics) has had some programs related to nanotechnology since 1992.

Germany

The Federal Ministry of Education, Science, Research, and Technology (BMBF) in Germany provides substantial national support for nanotechnology.

The Fraunhofer Institutes, Max Planck Institutes, and several universities have formed centers of excellence in the field. It is estimated that in 1997 BMBF supported programs on nanotechnology with a budget of approximately $50 million per year.

Two of the largest upcoming projects are "CESAR," a $50 million science centre in Bonn equally sponsored by the state and federal governments with about one-third of its research dedicated to nanoscience, and a new institute for carbon-reinforced materials near Karlsruhe ($4 million over 3 years, 1998-2001). BMBF is establishing five "centers of competence in nanotechnology" in Germany starting in 1998, with topics ranging from molecular architecture to ultraprecision manufacturing.

U.K.

A network programme (LINK Nanotechnology Programme) was launched in the United Kingdom in 1988 with an annual budget of about $2 million per year. The Engineering and Physical Sciences Research Council (EPSRC) is funding materials science projects related to nanotechnology with a total value of about $7 million for a five-year interval (1994-1999). About $1 million is specifically earmarked for nanoparticle research. The National Physical Laboratory established a forum called the National Initiative on Nanotechnology (NION) for promoting nanotechnology in universities, industry, and government laboratories.

France

The Centre National de la Récherche Scientifique (CNRS) has developed research programs on nanoparticles and nanostructured materials at about 40 physics laboratories and 20 chemistry laboratories in France. Synthesis methods include molecular beam and cluster deposition, lithography, electrochemistry, soft chemistry, and biosynthesis. Nanotechnology activity has grown within a wide variety of research groups, including ones focused on molecular electronics, large gap semiconductors and nanomagnetism, catalysts, nanofilters, therapy problems, agrochemistry, and even cements for ductile nanoconcretes.

It is estimated that CNRS spends about 2% of its budget and dedicates 500 researchers in 60 laboratories (or about $40 million per year) on projects related to nanoscience and nanotechnology. Companies collaborating to research and produce nanomaterials include Thompson, St. Gobain, Rhône Poulenc, Air Liquide, and IEMN. Also, there is the "French Club Nanotechnologie," aimed at promoting interactions in this field in France.

Sweden

The estimated total expenditure for research on nanotechnology in Sweden is ~ $10 million per year. There are four materials research consortia involved in this field: 1.

Ångström Consortium in Uppsala, with a budget of ~ $0.8 million per year in 1998 for surface nanocoatings 2. Nanometer Structures Consortium in Lund with a budget of ~ $3.5 million per year partially supported by ESPRIT (~ $1 million per year) 3. Cluster-based and Ultrafine Particle Materials in the University of Uppsala and Royal Institute of Technology with a budget of ~ $0.4 million per year in 1998 4. Brinell Centre at the Royal Institute of Technology

Switzerland

There is a Swiss national programme on nanotechnology with a special strength in instrumentation. The most advanced research centers are focused on nanoprobes and molecule manipulation on surfaces (IBM Research Laboratory in Zürich), devices and sensors (Paul Scherrer Institute), nanoelectronics (ETH Zürich), and self-assembling on surfaces in patterns determined by the substrate or template (L'École Polytechnique Fédérale de Lausanne).

The Netherlands

The most active research centers in the Netherlands are the DIMES institute at Delft University of Technology, which receives one-third support from industry, and the Philips Research Institute in Eindhoven, which researches self-assembling monolayers and patterning on metallic and silicon surfaces. The SST Netherlands Study Centre for Technology Trends is completing a study on nanotechnology and aims at promoting increased funding and research networking in the Netherlands.

Finland

The Academy of Finland and the Finnish Technology Development Centre began a three-year nanotechnology programme in 1997. The programme involves sixteen projects with funding of $9 million for a three-year period for nanobiology, functional nanostructures, nanoelectronics, and other areas. Research on actuators and sensors is the Finnish area of strength.

Belgium and Spain

Since about 1993 both Belgium and Spain have established nanotechnology programs, centers of excellence, and university-industry interactions.

Multinational Efforts

Large multinational companies with significant nanotechnology research activities in Europe include IBM (Zürich), Philips, Siemens, Bayer, and Hitachi. Degussa Co., with headquarters in Germany, is a commercial supplier since 1940 of microparticles and, now, nanoparticles. Western Europe has a variety of approaches to funding research on nanotechnology.

These are discussed in detail in other studies. IPTS (the Science and Technology Forecast Institute) has conducted a study on nanotechnology research in the EC. Other European studies published recently include those by VDI (1996), UNIDO (1997), the U.K. Parliamentary Office of Science and Technology (1997), and NANO network. The overall expenditure for nanotechnology research within the EC was estimated in 1997 to be over $128 million per year.

RUSSIA AND OTHER FSU COUNTRIES

Support for generation of nanoparticles and nanostructured materials has a tradition in Russia and other countries of the former Soviet Union (FSU) dating back to the mid-1970s; before 1990 an important part of this support was connected to defence research. The first public paper concerning the special properties of nanostructures was published in Russia in 1976. In 1979 the Council of the Academy of Sciences created a section on "Ultra- Dispersed Systems." Research strengths are in the areas of preparation processes of nanostructured materials and in several basic scientific aspects.

Metallurgical research for special metals, including those with nanocrystalline structures, has received particular attention; research for nanodevices has been relatively less developed. Due to funding limitations, characterization and utilization of nanoparticles and nanostructured materials requiring costly equipment are less advanced than processing. Russian government funds are allocated mainly for research personnel and less for infrastructure. Funding for nanotechnology is channeled via the Ministry of Science and Technology, the Russian Foundation for Fundamental Research, the Academy of Sciences, the Ministry of Higher Education, and other ministries with specific targets.

The Ministry of Higher Education has relatively little research funding. Overall, 2.8% of the civilian budget in Russia in 1997 was planned for allocation to science. There is no centralized programme on nanotechnology; however, there are components in specific institutional programs. Currently, about 20% of science research in Russia is funded via international organizations. The significant level of interest in the FSU can be identified by the relatively large participation at a series of Russian conferences on nanotechnology, the first in 1984, a second in 1989, and a third in 1993.

The Ministry of Science and Technology contributes to nanotechnology through several of its specific programs related to solid-state physics, surface science, fullerenes and nanostructures, and particularly "electronic and optical properties of nanostructures." This last programme involves a network of scientific centers: the Ioffe Institute in St. Petersburg, Lebedev Institute in Moscow, Moscow State University, Novgograd Institute of Microstructures, Novosibirsk Institute of Semiconductor Physics, and others. This research network has an annual meeting on nanostructures, physics, and technology, and has developed interactions with the PHANTOMS network in the EC. The U.S. Civilian Research and Development Foundation has provided research funds in the FSU for several projects related to nanotechnology, including "Highly Non-Equilibrium States and Processes in Nanomaterials" at the Ioffe Institute.

Russian government and international organizations are the primary research sponsors for nanotechnology in Russia. However, laboratories and companies privatized in the last few years, such as the Delta Research Institute in Moscow, are under development. With a relatively lower base in characterization and advanced computing, the research focus is on advanced processing and continuum modeling.

Research strengths are in the fields of physico-chemistry, nanostructured materials, nanoparticle generation and processing methods, and applications for hard materials, purification, and the oil industry, and biologically active systems. There are related programs in Ukraine, Belarus, and Georgia, mostly under the direction of the respective academies of sciences in these countries, that are dedicated to crystalline nanostructures and advanced structural and nanoelectronic materials. Several innovative processes, such as diamond powder production by detonation synthesis at SINTA in Belarus, are not well known abroad.

CLOSING REMARKS

Nanotechnology in the United States, Japan, and Western Europe is making progress in developing a suitable research infrastructure. The promise of nanotechnology is being realized through the confluence of advances in two fields:

- Scientific discovery that has enabled the atomic, molecular, and supramolecular control of material building blocks,
- Manufacturing that provides the means to assemble and utilize these tailored building blocks for new processes and devices in a wide variety of applications.

Technology programs cannot be developed without strong supporting science programs because of the scale and complexity of the nanosystems. The overlapping of discipline-oriented research with nanotechnology-targeted programs seems appropriate at this point in time. Highly interdisciplinary and multiapplication nanotechnology provides generic approaches that enable advances in other technologies, from dispersions, catalysts, and electronics to biomedicine. Essential trends include the following:

- Learning from nature (including templating, self-assembly, multifunctionality)

- Building up functional nanostructures from molecules
- Convergence of miniaturization and assembly techniques
- Novel materials by design
- Use of hierarchical/adaptive simulations

A characteristic of discovery in nanotechnology is the potential for revolutionary steps. The question "what if?" is progressively replaced by "at what cost?" The road from basic research to applications may vary from a few months to decades. Research and development is expensive, and the field needs support from related areas. The R&D environment should favour multiapplication and international partnerships. Based on the data for 1996 and 1997 collected during this WTEC study, 1997 government expenditures for nanotechnology research were at similar absolute levels in the United States, Japan, and Western Europe.

The largest funding opportunities for nanotechnology are provided by NSF in the United States (approximately $65 million per year for fundamental research), by MITI in Japan (approximately $50 million per year for fundamental research and development), and by BMBF in Germany (approximately $50 million per year for fundamental and applied research). Large companies in areas such as dispersions, electronics, multimedia, and bioengineering contribute to research to a larger extent in Japan and the United States than in Europe. While multinational companies are pursuing nanotechnology research activities in almost all developed countries, the presence of an active group of small and medium-size companies introducing new processes to the market is limited to the United States.

In the United States, individual and small-group researchers as well as industrial and national laboratories for specialized topics have established a strong position in synthesis and assembly of nanoscale building blocks and catalysts, and in polymeric and biological approaches to nanostructured materials. The Japanese large-group research institutes, and more recently academic laboratories, have made particular advances in nanodevices and nano-instrumentation. The European "mosaic" provides a diverse combination of university research, networks, and national laboratories with special performance in dispersion and coatings, nanobiotechnology, and nanoprobes. With a relatively lower base in characterization and computing infrastructure, the research focus in Russia is on physico-chemistry phenomena, advanced processing, and continuum modeling. Interest and economic support, particularly for device-related research, is growing in China, Australia, India, Taiwan, Korea, and Singapore.

Table. Government Expenditures on Nanotechnology Research in 1997, Based on the WTEC Site Interviews

Geographical Area	Annual Budged, NTR* ($ million)	Reletive Annual Budget NTR/GDP* (ppm)
Japan	120	27
United States	116	15
Western Europe	128	18
Other countries (FSU, China, Canada, Australia, Korea, Taiwan, Singapore)	70	-
Total	432	-

The pace of revolutionary discoveries that we are witnessing now in nanotechnology is expected to accelerate in the next decade worldwide. This will have a profound impact on existing and emerging technologies in almost all industry sectors, in conservation of materials and energy, in biomedicine, and in environmental sustainability.

Chapter 9

Global Technology Revolution

GENOMICS

By 2015, biotechnology will likely continue to improve and apply its ability to profile, copy, and manipulate the genetic basis of both plants and animal organisms, opening wide opportunities and implications for understanding existing organisms and engineering organisms with new properties. Research is even under way to create new free-living organisms, initially microbes with a minimal genome.

Genetic Profiling and DNA Analysis

DNA analysis machines and chip-based systems will likely accelerate the proliferation of genetic analysis capabilities, improve drug search, and enable biological sensors.

The genomes of plants (ranging from important food crops such as rice and corn to production plants such as pulp trees) and animals (ranging from bacteria such as *E. coli,* through insects and mammals) will likely continue to be decoded and profiled. To the extent that genes dictate function and behaviour, such extensive genetic profiling could provide an ability to better diagnose human health problems, design drugs tailored for individual problems and system reactions, better predict disease predispositions, and track disease movement and development across global populations, ethnic groups, and other genetic pools.

Note that a link between genes and function is generally accepted, but other factors such as the environment and phenotype play important modifying roles. Gene therapies will likely continue to be developed, although they may not mature by 2015.

Genetic profiling could also have a significant effect on security, policing, and law. DNA identification may complement existing biometric technologies (e.g., retina and fingerprint identification) for granting access to secure systems (e.g., computers, secured areas, or weapons), identifying criminals through DNA left at crime scenes, and authenticating items such as fine art. Genetic identification will likely become more commonplace tools in kidnapping, paternity, and fraud cases. Biosensors (some genetically engineered) may also aid in detecting biological warfare threats, improving food and water quality testing, continuous health monitoring, and medical laboratory analyses. Such capabilities could fundamentally change the way health services are rendered by greatly improving disease diagnosis, understanding predispositions, and improving monitoring capabilities.

Such profiling may be limited by technical difficulties in decoding some genomic segments and in understanding the implications of the genetic code. Our current technology can decode nearly all of the entire human gene sequence, but errors are still an issue, since Herculean efforts are required to decode the small amount of remaining sequences.

More important, although there is a strong connection between an organism's function and its genotype, we still have large gaps in understanding the intermediate steps in copying, transduction, isomer modulation, activation, immediate function, and this function's effect on larger systems in the organism. Proteomics (the study of protein function and genes) is the next big technological push after genomic decoding. Progress may likely rely on advances in bioinformatics, genetic code combination and sequencing (akin to hierarchical programming in computer languages), and other related information technologies.

Despite current optimism, a number of technical issues and hurdles could moderate genomics progress by 2015. Incomplete understanding of sequence coding, transduction, isomer modulation, activation, and resulting functions could form technological barriers to wide engineering successes. Extensive rights to own genetic codes may slow research and ultimately the benefits of the decoding. At the other extreme, the inability to secure patents from sequencing efforts may reduce commercial funding and thus slow research and resulting benefits.

In addition, investments in biotechnology have been cyclic in the past. As a result, advancements in research and development (R&D) may come in surges, especially in areas where the time to market (and thus time to return on investment) is long.

Cloning

Artificially producing genetically identical organisms through cloning will likely be significant for engineered crops, livestock, and research animals.

Cloning may become the dominant mechanism for rapidly bringing engineered traits to market, for continued maintenance of these traits, and for producing identical organisms for research and production. Research will likely continue on human cloning in unregulated parts of the world with possible success by 2015, but ethical and health concerns will likely limit wide-scale cloning of humans in regulated parts of the world. Individuals or even some states may also engage in human or animal cloning, but it is unclear what they may gain through such efforts.

Cloning, especially human cloning, has already generated significant controversies across the globe. Concerns include moral issues, the potential for errors and medical deficiencies of clones, questions of the ownership of good genes and genomes, and eugenics. Although some attempts at human cloning are possible by 2015, legal restrictions and public opinion may limit their extent. Fringe groups, however, may attempt human cloning in advance of legislative restrictions or may attempt cloning in unregulated countries.

Although expert opinions vary regarding the current feasibility of human cloning, at least some technical hurdles for human cloning will likely need to be addressed for safe, wide-scale use. "Attempts to clone mammals from single somatic cells are plagued by high frequencies of developmental abnormalities and lethality". Even cloned plant populations exhibit "substantial developmental and morphological irregularities". Research will need to address these abnormalities or at the very least mitigate their repercussions. Some believe, however, that human cloning may be accomplished soon if the research organization accepts the high lethality rate for the embryo and the potential generation of developmental abnormalities.

Genetically Modified Organisms

Beyond profiling genetic codes and cloning exact copies of organisms and microorganisms, biotechnologists can also manipulate the genetic code of plants and animals and will likely continue efforts to engineer certain properties into life forms for various reasons. Traditional techniques for genetic manipulation (such as cross-pollination, selective breeding, and irradiation) will likely continue to be extended by direct insertion, deletion, and modification of genes through laboratory techniques. Targets include food crops, production plants, insects, and animals.

Desirable properties could be genetically imparted to genetically engineered foods, potentially producing: improved taste; ultra-lean meats with reduced "bad" fats, salts, and chemicals; disease resistance; and artificially introduced nutrients (so-called "nutraceuticals"). Genetically modified organisms (GMOs) can potentially be engineered to improve their physical robustness, extend field and shelf life, tolerate herbicides, grow faster, or grow in previously unproductive environments (e.g., in high-salinity soils, with less water, or in colder climates).

Beyond systemic disease resistance, *in vivo* pesticide production has already been demonstrated (e.g., in corn) and could have a significant effect on pesticide production, application, regulation, and control with targeted release. Likewise, organisms could be engineered to produce or deliver

drugs for human disease control. Cow mammary glands might be engineered to produce pharmaceuticals and therapeutic organic compounds; other organisms could be engineered to produce or deliver therapeutics (e.g., the so-called "prescription banana"). If accepted by the population, such improved production and delivery mechanisms could extend the global production and availability of these therapeutics while providing easy oral delivery.

In addition to food production, plants may be engineered to improve growth, change their constitution, or artificially produce new products. Trees, for example, will likely be engineered to optimize their growth and tailor their structure for particular applications such as lumber, wood pulp for paper, fruiting, or carbon sequestering (to reduce global warming) while reducing waste byproducts. Plants might be engineered to produce bio-polymers (plastics) for engineering applications with lower pollution and without using oil reserves. Bio-fuel plants could be tailored to minimize polluting components while producing additives needed by the consuming equipment.

Genetic engineering of microorganisms has long been accepted and used. For example, *E. coli* has been used for mass production of insulin. Engineering of bacterial properties into plants and animals for disease resistance will likely occur.

Other animal manipulations could include modification of insects to impart desired behaviors, provide tagging (including GMO tagging), or prevent physical uptake properties to control pests in specific environments to improve agriculture and disease control.

Research on modifying human genes has already begun and will likely continue in a search for solutions to genetically based diseases. Although slowed by recent difficulties, gene therapy research will likely continue its search for useful mechanisms to address genetic deficiencies or for modulating physical processes such as beneficial protein production or control mechanisms for cancer. Advances in genetic profiling may improve our understanding and selection of therapy techniques and provide breakthroughs with significant health benefits. Some cloning of humans will be possible by 2015, but legal restrictions and public opinion may limit its actual extent. Controls are also likely for human modifications (e.g., clone-based eugenic modifications) for nondisease purposes. It is possible, however, that technology will enable genetic modifications for hereditary conditions (i.e., sickle cell anemia) through *in vitro* techniques or other mechanisms.

GMOs are also having a large effect on the scientific community as an enabling technology. Not only do "knock-out" animals (animals with selected DNA sequences removed from their genome) give scientists another tool to study the effect of the removed sequence on the animal, they also enable subsequent analysis of the interaction of those functions or components with the animal's entire system. Although knock-outs are not always complete, they provide another important tool to confirm or refute hypotheses regarding complex organisms.

Broader Issues and Implications

Extant capabilities in genomics have already created opportunities yet have generated a number of issues. As more organisms are decoded and the functional implications of genes are discovered, concerns about property and privacy rights for the sequencing will likely continue.

The ability to profile an individual's DNA is already raising concerns about privacy and excessive monitoring. Examples include databases of DNA signatures for use in criminal investigations, and the potential use of genetically based health predispositions by insurance companies or employers to deny coverage or to discriminate. The latter may raise policy issues regarding acceptable and unacceptable profiling for insurance or employment. This issue is further worrisome because the exact code-to-function mechanisms that trigger many disease predispositions are not well understood.

Issues may also arise if a strong genetic basis of human physical or cognitive ability is discovered. On the positive side, understanding a person's predisposition for certain abilities (or limitations) could enable custom educational or remediation programs that will help to compensate for genetic inclinations, especially in early years when their effect can be optimized. On the negative side, groups may use such analyses in arguments to discriminate against target populations (despite, for example, the fact that ethnic distribution variances of cognitive ability are currently believed to be wider than ethnic mean differences), aggravating social and international conflicts.

Although the genetic profiles of plants have been modified for centuries using traditional techniques, questions regarding the safety of genetically modified foods have sparked international concerns in the United Kingdom and Europe, forcing a campaign by biotechnology companies to argue the safety of the technology and its applications. Some have argued that genetic engineering is actually as safe or safer than traditional combinatorial techniques such as irradiated seeds, since there often is strong supporting information concerning the function of the inserted sequences.

Governments have been forced into the issue, resulting in education efforts, food labeling proposals, and heated international trade discussions between the United States and Europe on the importation of GMOs and their seedlings. As genetic modification becomes more common, it may become more difficult to label and separate GMOs, resulting in a forcing function to resolve the issue of how far the technology should be applied and whether separate markets can be maintained in a global economy. This debate is starting to have global effects as populations in other countries begin to notice the impassioned debates in the United Kingdom and Europe.

Some have likened the anti-biotechnology movement to the anti-nuclear-power movement in scope and tactics, although the low cost and wide availability of basic genomic equipment and know-how will likely allow practically any country, small business, or even individual to participate in genetic engineering. Such wide technology availability and low entry costs could make it impossible for any movement or government to control the spread and use of genomic technology. At an extreme, successful protest pressures on big biotechnology companies together with wide technology availability could ultimately drive genomic engineering "underground" to groups outside such pressures and outside regulatory controls that help ensure safe and ethical uses. This could ironically facilitate the very problems that the anti-biotechnology movement is hoping to prevent.

Cloning and genetic modification also raise biodiversity concerns. Standardization of crops and livestock have already increased food supply vulnerabilities to diseases that can wipe out larger areas of production. Genetic modification may increase our ability to engineer responses to these threats, but the losses may still be felt in the production year unless broad-spectrum defenses are developed.

In addition to food safety, the ability to modify biological organisms holds the possibility of engineered biological weapons that circumvent current or planned countermeasures. On the other hand, genomics could aid in biological warfare defence (e.g., through improved understanding and control of biological function both in and between pathogens and target hosts as well as improved capability for engineered biosensors). Advances in genomics, therefore, could advance a race between threat engineering and countermeasures. Thus, although genetic manipulation is likely to result in medical advances, it is unclear whether we will be in a safer position in the future.

The rate at which GMO benefits are felt in poorer countries may depend on the costs of using patented organisms, marketing demands and approaches, and the rate at which crops become ubiquitous and inseparable from unmodified strains. Consider, for example, current issues related to human immunodeficiency virus (HIV) drug development and dissemination in poorer countries. Patentability has fueled research investments, but many poorer countries with dire needs cannot afford the latest drugs and must wait for handouts or patent expiration. Globalization, however, may fuel dissemination as multi-national companies invest in food production across the globe. Also, the rewards from opening previously unproductive land for production may provide the financial incentive to pay the premium for GMOs. Furthermore, widely available genomic technology could allow academics, nonprofit small businesses, and developing countries to develop GMOs to alleviate problems in poorer regions; larger biotechnology companies will focus on markets requiring capital-intensive R&D.

Finally, moral issues may play a large role in modulating the global effect of genomics trends. Some people simply believe it is improper to engineer or modify biological organisms using the new techniques. Unplanned side effects (e.g., the imposition of arthritis in current genetically modified pigs) will likely support such opposition. Others are concerned with the real danger of eugenics programs or of the engineering of dangerous biological organisms.

THERAPIES AND DRUG DEVELOPMENT

Technology

Beyond genetics, biotechnology will likely continue to improve therapies for preventing and treating disease and infection. New approaches might block a pathogen's ability to enter or travel in the body, leverage pathogen vulnerabilities, develop new countermeasure delivery mechanisms, or modulate or augment the immune response to recognizing new pathogens. These therapies may counter the current trend of increasing resistance to extant antibiotics, reshaping the war on infections.

In addition to addressing traditional viral and bacterial problems, therapies are being developed for chemical imbalances and modulation of chemical stasis. For example, antibodies are being developed that attack cocaine in the body and may be used to control addiction. Such approaches could have a significant effect on modifying the economics of the global illegal drug trade while improving conditions for users.

Drug development will likely be aided by various technology trends and enablers. Computer simulations combined with proliferating trends for molecular imaging technologies (e.g., atomic-force microscopes, mass spectroscopy, and scanning probe microscopes) may continue to improve our ability to design molecules with desired functional properties that target specific receptors, binding sites, or markers, complementing combinatorial drug search with rational drug design. Simulations of drug interactions with target biological systems could become increasing useful in understanding drug efficacy and safety.

For example, Dennis Noble's complex virtual heart simulation has already contributed to U.S. Food and Drug Administration (FDA) approval of a cardiac drug by helping to understand the mechanisms and significance of an effect noticed in the clinical trial. For some better understood systems such as the heart, this approach may become a dominant complement to clinical drug trials by 2015, whereas other more complex systems (e.g., the brain) will likely require more research on the system function and biology.

Broader Issues and Implications

R&D costs for drug development are currently extremely high and may even be unsustainable, with averages of approximately $600 million per drug brought to market. These costs may drive the pharmaceutical industry to invest heavily in technology advances with the goal of long-term viability of the industry. Combined with genetic profiling, drug development tailored to genotypes, chemical simulation and engineering programs, and drug testing simulations may begin to change pharmaceutical development from a broad application trial-and-error approach to custom drug development, testing, and prescription based on a deeper understanding of subpopulation response to drugs.

This understanding may also rescue drugs previously rejected because of adverse reactions in small populations of clinical trials. Along with the potential for improving success rates, reducing trial costs, and opening new markets for narrowly targeted drugs, tailoring drugs to subpopulations will also have the opposite effect of reducing the size of the market for each drug. Thus, the economics of the pharmaceutical and health industries will likely change significantly if these trends come to fruition. Note that patent protection is not uniformly enforced across the globe for the pharmaceutical industry. As a result, certain regions (e.g., Asia) may continue to focus on production of non-legacy (generic) drugs, and other regions will likely continue to pursue new drugs in addition to such low-margin pharmaceuticals.

BIOMEDICAL ENGINEERING

Multidisciplinary teaming is accelerating advances and products in biomedical engineering and technology of organic and artificial tissues, organs, and materials.

Organic Tissues and Organs

Advances in tissue and organ engineering and repair are likely to result in organic and artificial replacement parts for humans. New advances in tissue regeneration and repair continue to improve our ability to resolve health problems within our bodies.

The field of tissue engineering, which is barely a decade old, has already led to engineered commercial skin products for wound treatment. Growth of cartilage for repair and replacement is at the stage of clinical testing, and treatment of heart disease via growth of functional tissue by 2015 is a realistic goal. These advances will depend upon improved biocompatible (or bioabsorbable) scaffold materials, development of 3D vascularized tissues and multicellular tissues, and an improved understanding of the *in vivo* growth process of cellular material on such scaffolds.

Research and applications of stem cell therapies will likely continue and expand, using these unspecialized human cells to augment or replace brain or body functions, organs (e.g., heart, kidney, liver, pancreas), and structures. As the most unspecialized stem cells are found in early stage embryos or fetal tissue, an ethical debate is ensuing regarding the use of stem cells for research and therapy. Alternatives such as the use of adult human stem cells or stem cell culturing may ultimately produce large-scale cell supplies with reduced ethical concerns. Current debates have limited U.S.

government funding for stem cell research, but the potential has attracted substantial private funding.

Xenotransplantations (transplantation of body parts from one species to a different species) could be improved, aided by attempts to genetically modify donor tissue and organ antibodies, complements, and regulatory proteins to reduce or eliminate rejection. Baboons or pigs, for example, may be genetically modified and cloned to produce organs for human transplant, although large-scale success may not occur by 2015.

Beyond rejection, the significance of xenotransplants is likely to be modulated by concerns that diseases such as retro viruses might jump from animals to people as a result of the transplantation techniques. Ethical (e.g., animal rights) and moral concerns as well as possible patenting issues may also result in regulations and limitations on xenotransplants, limiting their significance.

Artificial Materials, Organs, and Bionics

In addition to organic structures, advances are likely to continue in engineering artificial tissues and organs for humans.

Multi-functional materials are being developed that provide both structure and function or that have different properties on different sides, enabling new applications and capabilities. For example, polymers with a hydrophilic shell around a hydrophobic core (biomimetic of micelles) can be used for timed release of hydrophobic drug molecules, as carriers for gene therapy or immobilized enzymes, or as artificial tissues. Sterically stabilized polymers could also be used for drug delivery.

Other materials are being developed for various biomedical applications. Fluorinated colloids, for example, are being developed that take advantage of the high electronegativity of fluorine to enhance *in vivo* oxygen transport (as a blood substitute during surgery) and for drug delivery. Hydrogels with controlled swelling behaviour are being developed for drug delivery or as templates to attach growth materials for tissue engineering. Ceramics such as bioactive calcia-phosphate-silica glasses (gel-glasses), hydroxyapetite, and calcium phosphates can serve as templates for bone growth and regeneration.

Bioactive polymers (e.g., polypeptides) can be applied as meshes, sponges, foams, or hydrogels to stimulate tissue growth. Coatings and surface treatments are being developed to increase biocompatibility of implanted materials (for example, to overcome the lack of endothelial cells in artificial blood vessels and reduce thrombosis). Blood substitutes may change the blood storage and retrieval systems while improving safety from blood-borne infections.

New manufacturing techniques and information technology are also enabling the production of biomedical structures with custom sizing and shape. For example, it may become commonplace to manufacture custom ceramic replacement bones for injured hands, feet, and skull parts by combining computer tomography and "rapid prototyping" to reverse engineer new bones layer by layer.

Beyond structures and organs, neural and sensor prosthetics could begin to become significant by 2015. Retinas and cochlear implants, bypasses of spinal and other nerve damage, and other artificial communications and stimulations may improve and become more commonplace and affordable, eliminating many occurrences of blindness and deafness. This could eliminate or reduce the effect of serious handicaps and change society's response from accommodation to remediation.

Biomimetics and Applied Biology

Recent techniques such as functional brain imaging and knock-out animals are revolutionizing our endeavors to understand human and animal intelligence and capabilities. These efforts should, by 2015, make significant inroads in improving our understanding of phenomena such as false memories, attention, recognition, and information processing, with implications for better understanding people and designing and interfacing artificial systems such as autonomous robots and information systems.

Neuromorphic engineering (which bases its architecture and design principles on those of biological nervous systems) has already produced novel control algorithms, vision chips, head-eye systems, and biomimetic autonomous robots. Although not likely to produce systems with wide intelligence or capabilities similar to those of higher organisms, this trend may produce systems by 2015 that can robustly perform useful functions such as vacuuming a house, detecting mines, or conducting autonomous search.

Surgical and Diagnostic Biotechnology

Biotechnology and materials advances are likely to continue producing revolutionary surgical procedures and systems that will significantly reduce hospital stays and cost and increase effectiveness. New surgical tools and techniques and new materials and designs for vesicle and tissue support will likely continue to reduce surgical invasiveness and offer new solutions to medical problems. Techniques such as angioplasty may continue to eliminate whole classes of surgeries; others such as laser perforations of heart tissue could promote regeneration and healing.

Advances in laser surgery could refine techniques and improve human capability eye surgery to replace glasses), especially as costs are reduced and experience spreads. Hybrid imaging techniques will likely improve diagnosis, guide human and robotic surgery, and aid in basic understanding of body and brain function. Finally, collaborative information technology (e.g., "telemedicine") will likely extend specialized medical care to remote areas and aid in the global dissemination of medical quality and new advances.

Broader Issues and Implications

By 2015, one can envision: effective localized, targeted, and controlled drug delivery systems; long-lived implants and prosthetics; and artificial skin, bone, and perhaps heart muscle or even nerve tissue. A host of social, political, and ethical issues such as those discussed above will likely accompany these developments.

Biomedical advances (combined with other health improvements) are already increasing human life span in countries where they are applied. New advances by 2015 are likely to continue this trend, accentuating issues such as shifts in population age demographics, financial support for retired persons, and increased health care costs for individuals. Advances, however, may improve not only life expectancy but productivity and utility of these individuals, offsetting or even overcoming the resulting issues.

Many costly and specialized medical techniques are likely to initially benefit citizens who can afford better medical care (especially in developed countries, for example); wider global effects may occur later as a result of traditional trickle-down effects in medicine. Some technologies (e.g., telemedicine) may have the opposite trend where low-cost technologies may enable cost-effective consulting with specialists regardless of location. However, access to technology may greatly mediate this dispersal mechanism and may place additional demands on technology upgrades and education. Countries that remain behind in terms of technological infrastructures may miss many of these benefits.

Theological debates have also raised concerns about the definition of what constitutes a human being, since animals are being modified to produce human organs for later xenotransplantation in humans. Genetic profiling may help to inform this debate as we understand the genetic differences between humans and animals.

Improved understanding of human intelligence and cognitive function could have broader legal and social effects. For example, an understanding of false memories and how they are created could have an effect on legal liabilities and courtroom testimony. Understanding innate personal capabilities and job performance requirements could help us determine who would make better fighter pilots, who has an edge in analyzing complex images, and what types of improved training could improve people's capabilities to meet the special demands of their chosen careers. Ethical concerns could arise concerning discrimination against people who lack certain innate skills, requiring objective and careful measures for hiring and promotion.

Eventually, neural and sensory implants (combined with trends toward pervasive sensors in the environment and increased information availability) could radically change the way people sense, perceive, and interact with natural and artificial environments. Ultimately, these new capabilities could create new jobs and functions for people in these environments. Such innovations may first develop for individuals with particularly challenging and critical functions (e.g., soldiers, pilots, and controllers), but innovations may first develop in other quarters (e.g., for entertainment or business functions), given recent trends. Initial research indicates the feasibility of such implants and interactions, but it is unclear whether R&D and investments will accelerate enough to realise even such early applications by 2015. Current trends have concentrated on medical prosthetics where research prototypes are already appearing so it appears likely that globally significant systems will appear in this domain first.

THE PROCESS OF MATERIALS ENGINEERING

New materials can often be critical enabling drivers for new systems and applications with significant effects. However, it may not be obvious how enabling materials affect more observable trends and applications. A common process model from materials engineering can help to show how materials appear likely to break previous barriers in the process that ultimately results in applications with potential global benefits.

Developments in materials science and engineering result from interdisciplinary materials research. This development can be conveniently represented by the schematic description of the materials engineering process from concept to product/ application. This process view is a common approach in materials research circles and similar representations may be found in the literature. Current trends in materials research that could result in global effects by 2015 are categorized below according to the process description of Figure provides an example of the development process in the area of electroactive polymers for robotic devices and artificial muscles.

Concept/Materials Design

Biomimetics is the design of systems, materials, and their functionality to mimic nature. Current examples include layering of materials to achieve the hardness of an abalone shell or trying to understand why spider silk is stronger than steel. *Combinatorial materials design* uses computing power (sometimes together with massive parallel experimentation) to screen many different materials possibilities to optimize properties for specific applications (e.g., catalysts, drugs, optical materials).

Materials Selection, Preparation, and Fabrication

Composites are combinations of metals, ceramics, polymers, and biological materials that allow multi-functional behaviour. One common practice is reinforcing polymers or ceramics with ceramic fibers to increase strength while retaining light weight and avoiding the brittleness of the monolithic ceramic. Materials used in the body often combine biological and structural functions (e.g., the encapsulation of drugs).

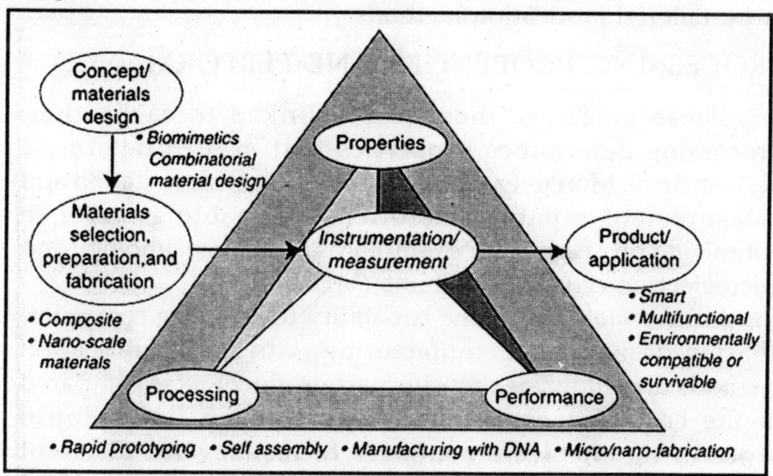

Fig. The General Materials Engineering Process

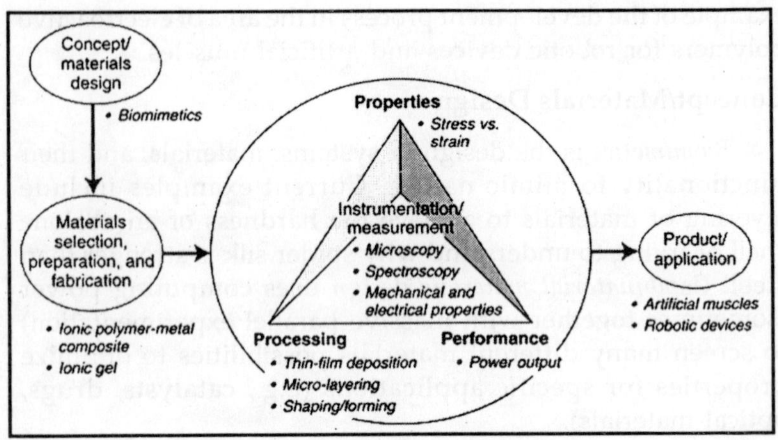

Fig. Materials Engineering Process
Applied to Electroactive Polymers

Nanoscale materials, i.e., materials with properties that can be controlled at submicrometer (<10^{-6} m) or nanometer (10^{-9} m) level, are an increasingly active area of research because properties in these size regimes are often fundamentally different from those of ordinary materials. Examples include carbon nanotubes, quantum dots, and biological molecules. These materials can be prepared either by purification methods or by tailored fabrication methods.

PROCESSING, PROPERTIES, AND PERFORMANCE

These areas are inextricably linked to each other: Processing determines properties that in turn determine performance. Moreover, the sensitivity of instrumentation and measurement capability is often the enabling factor in optimizing processing, for example, as for nanotechnology and microelectromechanical systems (MEMS).

Rapid prototyping is the capability to combine computer-assisted design and manufacturing with rapid fabrication methods that allow inexpensive part production (as compared to the cost of a conventional production line). Rapid prototyping enables a company to test several different inexpensive prototypes before committing infrastructure investments to an approach. Combined with manufacturing system improvements to allow flexibility of approach and machinery, rapid prototyping can lead to an *agile manufacturing* capability.

Alternatively, the company can use its virtual capability to design and then outsource product manufacturing, thus offloading capital investment and risk. This capability is synergistic with the information technology revolution in the sense that it is a further factor in globalizing manufacturing capability and enabling organizations with less capital to have a significant technological effect. For the Department of Defence (DoD), it could reduce or eliminate requirements for warehousing large amounts of spares and, for example, could enable the Air Force to "fly before they buy."

Self-assembly refers to the use in materials processing or fabrication of the tendency of some materials to organize themselves into ordered arrays (e.g., colloidal suspensions). This provides a means to achieve structured materials "from the bottom up" as opposed to using manufacturing or fabrication methods such as lithography, which is limited by the measurement and instrumentation capabilities of the day. For example, organic polymers have been tagged with dye molecules to form arrays with lattice spacing in the visible optical wavelength range and that can be changed through chemical means. This provides a material that fluoresces and changes colour to indicate the presence of chemical species.

Manufacturing with DNA might represent the ultimate biomimetic manufacturing scheme. It consists of "functionalizing small inorganic building blocks with DNA and then using the molecular recognition processes associated with DNA to guide the assembly of those particles or building blocks into extended structures". Using this approach, Mirkin and colleagues demonstrated a highly selective and sensitive DNA-based chemical assay method using 13 nm diameter gold particles with attached DNA sequences. This approach is compatible with the commonly used polymerase chain reaction (PCR) method of amplification of the amount of the target substance.

Micro- and nano-fabrication methods include, for example, lithography of coupled micro- or nano-scale devices on the same semiconductor or biological material. It is important to note the crucial role played in the development of these techniques by the parallel development of instrumentation and measurement devices such as the Atomic Force Microscope (AFM) and the various Scanning Probe Microscopes (SPMs).

Product/Application

The trends described above will likely work in concert to provide materials engineers with the capability to design and produce advanced materials that will be:

- *Smart*—Reactive materials combining sensors and actuators, perhaps together with computers, to enable response to environmental conditions and changes thereof. (Note, however, that limitations include the sensitivity of sensors, the performance of actuators, and the availability of power sources with required magnitude compatible with the desired size of the system.) An example might be robots that mimic insects or birds for applications such as space exploration, hazardous materials location and treatment, and unmanned aerial vehicles (UAVs).
- *Multi-functional*—MEMS and the "lab-on-a-chip" are excellent examples of systems that combine several functions. Another example is a drug delivery system using a hydrogel with hydrophilic exterior and hydrophobic interior. Consider also aircraft skins fabricated from radar-absorbing materials that incorporate avionic links and the ability to modify shape in response to airflow.
- *Environmentally compatible or survivable*—The development of composite materials and the ability to tailor materials at the atomic level will likely provide opportunities to make materials more compatible with the environments in which they will be used. Examples might include prosthetic devices that serve as templates for the growth of natural

tissue and structural materials that strengthen during service.

SMART MATERIALS

Technology

Several different types of materials exhibit sensing and actuation capabilities, including ferroelectrics (exhibiting strain in response to a electric field), shape-memory alloys (exhibiting phase transition-driven shape change in response to temperature change), and magnetostrictive materials (exhibiting strain in response to a magnetic field). These effects also work in reverse, so that these materials, separately or together, can be used to combine sensing and actuation in response to environmental conditions. They are currently in widespread use in applications from ink-jet printers to magnetic disk drives to anti-coagulant devices.

An important class of smart materials is composites based upon lead zirconate titanate (PZT) and related ferroelectric materials that allow increased sensitivity, multiple frequency response, and variable frequency. An example is the "Moonie"—a PZT transducer placed inside a half-moon-shaped cavity, which provides substantial amplification of the response. Another example is the use of composites of barium strontium titanate and non-ferroelectric materials that provide frequency-agile and field-agile responses. Applications include sensors and actuators that can change their frequency either to match a signal or to encode a signal. Ferroelectrics are already in use as nonvolatile memory elements for smart cards and as active elements in smart skis that change shape in response to stress.

Another important class of materials is smart polymers (e.g., ionic gels that deform in response to electric fields). Such electro-active polymers have already been used to make "artificial muscles". Currently available materials have limited mechanical power, but this is an active research area with potential applications to robots for space exploration, hazardous duty of various types, and surveillance. Hydrogels that swell and shrink in response to changes in pH or temperature are another possibility; these hydrogels could be used to deliver encapsulated drugs in response to changes in body chemistry (e.g., insulin delivery based upon glucose concentration). Another variation on this trend for controlled release of drugs is materials with hydrophilic exterior and hydrophobic interior.

Broader Issues and Implications

A world with pervasive, networked sensors and actuators (e.g., on and part of walls, clothing, appliances, vehicles, and the environment) promises to improve, optimize, and customise the capability of systems and devices through availability of information and more direct actuation. Continuously available communication capability, ability to catalog and locate tagged personal items, and coordination of support functions have been espoused as benefits that may begin to be realized by 2015.

The continued development of small, low-profile biometric sensors, coupled with research on voice, handwriting, and fingerprint recognition, could provide effective personal security systems. These could be used for identification by police/military and also in business, personal, and leisure applications.

Combined with today's information technologies, such uses could help resolve nagging security and privacy concerns while enabling other applications such as improved handgun safety (through owner identification locks) and vehicle theft control.

Other potential applications of smart materials that would be enabled by 2015 include: clothes that respond to weather, interface with information systems, monitor vital signs, deliver medicines, and automatically protect wounds; airfoils that respond to airflow; buildings that adjust to the weather; bridges and roads that sense and repair cracks; kitchens that cook with wireless instructions; virtual reality telephones and entertainment centers; and personal medical diagnostics (perhaps interfaced directly with medical care centers). The level of development and integration of these technologies into everyday life will probably depend more on consumer attitudes than on technical developments.

In addition to the surveillance and identification functions mentioned under smart materials above, developments in robotics may provide new and more sensitive capabilities for detecting and destroying explosives and contraband materials and for operating in hazardous environments.

Increases in materials performance, both for power sources and for sensing and actuation, as well as integration of these functions with computing power, could enable these applications.

Such trend potentials are not without issues. Pervasive sensory information and access to collected data raise significant privacy concerns. Also, the pace of development will likely depend on investment levels and market drivers. In many cases the immediate benefits and cost savings from smart material applications will continue to drive development, but more exotic materials research may depend on public commitment to research and belief in investing in longer-term rewards.

SELF-ASSEMBLY

Technology

Examples of self-assembling materials include colloidal crystal arrays with mesoscale (50-500 nm) lattice constants that form optical diffraction gratings, and thus change colour as the array swells in response to heat or chemical changes. In the case of a hydrogel with an attached side group that has molecular recognition capability, this is a chemical sensor. Self-assembling colloidal suspensions have been used to form a light-emitting diode (nanoscale), a porous metal array (by deposition followed by removal of the colloidal substrate), and a molecular computer switch.

The DNA-based self-assembly mentioned above was achieved by attaching non-linking DNA strands to metal nanoparticles and adding a linking agent to form a DNA lattice. This can be turned into a biosensor or a nanolithography technique for biomolecules.

Broader Issues and Implications

Development of self-assembly methods could ultimately provide a challenge to top-down lithography approaches and molecular manufacturing approaches. As a result, it could define the next manufacturing methodology at some time beyond 2015.

For example, will self-assembly methods "trump" lithography (the miracle technology of the semiconductor revolution) over the next decade or two?

RAPID PROTOTYPING

Technology

This manufacturing approach integrates computer-aided design (CAD) with rapid forming techniques to rapidly create a prototype (sometimes with embedded sensors) that can be used to visualize or test the part before making the investment in tooling required for a production run. Originally, the prototypes were made of plastic or ceramic materials and were not functional models, but now the capability exists to make a functional part, e.g., out of titanium.

Broader Issues and Implications

As discussed above, agile manufacturing systems are envisioned that can connect the customer to the product throughout its life cycle and enable global business enterprises. An order would be processed using a computer-aided design, the manufacturing system would be configured in real time for the specific product (e.g., model, style, colour, and options), raw materials and components would be acquired just in time, and the product would be delivered and tracked throughout its life cycle (including maintenance and recycling with identification of the customer). Components of the business enterprise could be dynamically based in the most cost-effective locations with all networked together globally. The growth of this type of business enterprise could accelerate business globalization.

BUILDINGS

Research on composite materials, waste management, and recycling has reached the stage where it is now feasible to construct buildings using materials fabricated from significant amounts of indigenous waste or recycled material content. These approaches are finding an increasing number of cost-effective applications, especially in developing countries. Examples include the Petronas Twin Towers in Kuala Lumpur, Malaysia.

These towers are the tallest buildings on earth and are made with reinforced concrete rather than steel. A roofing material used in India is made of natural fibre and agro-industrial waste. Prefabricated composite materials for home construction have also been developed in the United States, and a firm in the Netherlands is developing a potentially ubiquitous, inexpensive housing approach targeted for developing countries that uses spray-forming over an inflatable air shell.

TRANSPORTATION

An important trend in transportation is the development of lightweight materials for automobiles that increase energy efficiency while reducing emissions. Here the key issue is the strength-to-weight ratio versus cost. Advanced composites with polymer, metal, or ceramic matrix and ceramic reinforcement are already in use in space systems and aircraft. These composites are too expensive for automobile applications, so aluminum alloys are being developed and introduced in cars such as the Honda Insight, the Audi A8 and AL2, and the GM EV1.

Although innovation in both design and manufacturing is needed before such all-aluminum structures can become widespread, aluminum content in luxury cars and light trucks has increased in recent years. Polymer matrix, carbon-fibre (C-fibre) reinforced composites could enable high mileage cars, but C-fibre is currently several times more expensive than steel. Research sponsored by the Department of Energy (DOE) at Oak Ridge National Laboratory is working to develop cheap C-fibers which could have wider application and effect.

Spurred by California's regulations concerning ultra-low-emission vehicles, both Honda and Toyota have introduced gasoline-electric hybrid vehicles. The U.S. government and industry consortium called Partnership for a New Generation of Vehicles (PNGV) has demonstrated prototype hybrid vehicles that use both diesel/electric and diesel/fuel-cell power plants and has established 2008 as the goal for a production vehicle. These vehicles use currently available materials, but the cost reduction issues described above will be critical in bringing production costs to levels that will allow significant market penetration.

ENERGY SYSTEMS

If the ready availability of oil continues, it may be difficult for technology trends to be much of a driving force in global energy between now and 2015. Key questions have to do with continued oil imports, continued use of coal, sources of natural gas, and the fate of nuclear power. Nevertheless, technology may have significant effects in some areas.

Along with investments in solar energy, current investments in battery technology and fuel cells could enable continued trends in more portable devices and systems while extending operating times.

Developments in materials science and engineering may enable the energy systems of 2015 to be more distributed with a greater capability for energy storage, as well as energy system command, control, and communication. High-temperature superconducting cables, transformers, and storage devices could begin to increase energy transmission and distribution capabilities and power quality in this time frame.

The continued development of renewable energy could be enhanced by the combination of cheap, lightweight, recyclable materials (and perhaps the genetic engineering of biomass fuels) to provide cost-effective energy for developing countries without existing, well-developed energy infrastructures as well as for remote locations.

Significant changes in developed countries, however, may be driven more by existing social, political, and business forces, since the fuel mix of 2015 will still be strongly based on fossil fuels. Environmental concerns such as global warming and pollution might shift this direction, but it would likely require long-term economic problems (e.g., a prolonged rise in the price of oil) or distribution problems (e.g., supplies interrupted by military conflicts) to drive advances in renewable energy development.

NEW MATERIALS

Materials research may provide improvements in properties by 2015 in a number of additional areas, leading to significant effects.

SiC, GaN, and other wide band gap semiconductors are being investigated as materials for high-power electronics.

Functionally graded materials (i.e., materials whose properties change gradually from one end to the other) can form useful interlayers between mechanically, thermally, or electrically diverse components.

Anodes, cathodes, and electrolytes with higher capacity and longer lifetime are being developed for improved batteries and fuel cells.

High-temperature (ceramic) superconductors discovered in 1986 can currently operate at liquid nitrogen (rather than liquid helium) temperatures. Prototype devices such as electrical transmission cables, transformers, storage devices, motors, and fault current limiters have now been built and demonstrated. Niche application on electric utility systems should begin by 2015 (e.g., replacement of underground cables in cities and replacement of older substation transformers).

Nonlinear optical materials such as doped $LiNbO_3$ are being investigated for ultraviolet lasers (e.g., to enable finer lithography). Efforts are under way to increase damage threshold and conversion efficiency, minimize divergence, and tailor the absorption edge. Hard materials such as nanocrystalline coatings and diamonds are being developed for applications such as computer disk drives and drill bits for oil and gas exploration, respectively.

High-temperature materials such as ductile intermetallics and ceramic matrix composites are being developed for aerospace applications and for high-efficiency energy and petrochemical conversion systems.

NANOMATERIALS

This area combines nanotechnology and many applications of nanostructured materials. One important research area is the formation of semiconductor "quantum dots" (i.e., several nanometer-size, faceted crystals) by injecting precursor materials conventionally used for chemical-vapour deposition of semiconductors into a hot liquid surfactant.

This "quantum dot" is in reality a macromolecule because it is coated with a monolayer of the surfactant, preventing agglomeration. These materials photoluminesce at different frequencies (colors) depending upon their size, allowing optical multiplexing in biological labeling.

Another important class of nanomaterials is nanotubes (the open cylindrical sisters of fullerenes). Possible applications are field-emission displays (Mitsubishi research), nanoscale wires for batteries, storage of Li or H_2, and thermal management (heat pipes or insulation—the latter taking advantage of the anisotropy of thermal conductivity along and perpendicular to the tube axis).

Another possibility is to use nanotubes (or fibers built from them) as reinforcement for composite materials. Presumably because of the nature of the bonding, it is predicted that nanotube-based material could be 50 to 100 times stronger than steel at one-sixth of the weight if current technical barriers can be overcome.

Nanoscale structures with desirable mechanical and other properties may also be obtained through processing. Examples include strengthening of alloys with nanoscale grain structure, increased ductility of metals with multi-phase nanoscale microstructure, and increased flame retardancy of plastic nanocomposites.

NANOTECHNOLOGY

Much has been made of the trend toward producing devices with ever-decreasing scale. Many people have projected that nanometer-scale devices will continue this trend, bringing it to unprecedented levels. This includes scale reduction not only in microelectronics but also in fields such as MEMS and quantum-switch-based computing in the shorter term. These advances have the potential to change the way we engineer our environment, construct and control systems, and interact in society.

Nanofabricated Computation Devices

Nanofabricated Chips.

SEMATECH—the leading industry group in the semiconductor manufacturing business—is calling for the development of nanoscale semiconductors in their latest International Technology Roadmap for Semiconductors (ITRS). The roadmap calls for a 35 nm gate length in 2015 with a total number of functions in high-volume production microprocessors of around 4.3 billion. For low-volume, high-performance processors, the number of functions may approach 20 billion. Corresponding memory chips (DRAMs) are targeted to hold around 64 gigabytes. These roadmap targets would continue the exponential trend in processing power, fueling advances in information technology. Although a number of engineering challenges exist (such as lithography, interconnects, and defect management), obstacles to achieve at least this level of performance do not seem insurmountable.

Given unforeseen shortfalls in the economic production of these chips (e.g., because of very high manufacturing costs or unacceptably large numbers of manufacturing defects), several alternatives seem possible. Defect-tolerant computer architectures such as those prototyped on a small scale by Hewlett-Packard offer one alternative. These alternative methods provide some level of additional robustness to the performance goals set by the ITRS.

However, in the years following 2015, additional difficulties will likely be encountered, some of which may pose serious challenges to traditional semiconductor manufacturing techniques. In particular, limits to the degree that interconnections or "wires" between transistors may be scaled could in turn limit the effective computation speed of devices because of materials properties and compatibility, despite incremental present-day advances in these areas. Thermal dissipation in chips with extremely high device densities will also pose a serious challenge. This issue is not so much a fundamental limitation as it is an economic consideration, in that heat dissipation mechanisms and cooling technology may be required that add to total system cost, thereby adversely affecting marginal cost per computational function for these devices.

Quantum-Switch-Based Computing.

One potential long-term solution for overcoming obstacles to increased computational power is computing based on devices that take advantage of various quantum effects. The core innovation in this work is the use of quantum effects, such as spin polarization of electrons, to determine the state of individual switches. This is in contrast to more traditional microelectronics, which are based on macroscopic properties of large numbers of electrons, taking advantage of materials properties of semiconductors.

Various concepts of quantum computers are attractive because of their massive parallelism in computation, but they are not anticipated to have significant effects by 2015. These concepts are qualitatively different from those employed in traditional computers and will hence require new computer architectures. The types of computations (and hence applications) that can be quickly performed using these computers are not the same as those readily addressed by today's digital computer.

Several workers in the area have devised algorithms for

problems that are very computationally intensive (and thus time-consuming) for existing digital computers, which could be made much faster using the physics of quantum computers. Examples of these problems include factoring large numbers (essential for cryptographic applications), searching large databases, pattern matching, and simulation of molecular and quantum phenomena.

A preliminary survey of work in this area indicates that quantum switches are unlikely to overcome major technical obstacles, such as error correction, de-coherence and signal input/output, within the next 15 years. If this were indeed the case, quantum-switch-based computing does not appear to be competitive with traditional digital electronic computers within the 2015 timeframe.

Bio-Molecular Devices and Molecular Electronics

Many of the same manufacturing and architectural challenges discussed above regarding quantum computing also hold true for molecular electronics. Molecular electronic devices could operate as logic switches through chemical means, using synthesized organic compounds. These devices can be assembled chemically in large numbers and organized to form a computer. The main advantage of this approach is significantly lower power consumption by individual devices.

Several approaches for such devices have been devised, and experiments have shown evidence of switching behaviour for individual devices. Several research groups have proposed interconnection between devices using carbon nanotubes, which provide high conductivity using single molecular strands of carbon. Progress has been made toward raising the operating temperature of these switches to nearly room temperature, making the switching process reversible, and increasing the overall amount of current that can be switched using these devices.

Several major outstanding issues remain with respect to molecular electronics. One issue is that molecular memories must be able to maintain their state, just as in a digital electronic computer. Also, given that the manufacturing and assembly process for these devices will lead to device defects, a defect-tolerant computer architecture needs to be developed. Fabricating reliable interconnects between devices using carbon nanotubes (or some other technology) is an additional challenge.

A significant amount of work is ongoing in each of these areas. Even though experimental progress to date in this area has been substantial, it seems unlikely (as with quantum computing) that molecular computers could be developed within the next 15 years that would be relatively attractive (from a price/performance standpoint) compared with conventional electronic computers.

Broader Issues and Implications

Examining the potential for developing qualitatively different computational capabilities from different technology bases is a challenging exercise. The history of computing over the last 50 years has seen one major shift in technology base (from vacuum tubes to semiconductor transistors), with a corresponding shift not just in computational power but also in attitudes about the value of computers. Ideas of computers as simple machines for computation gave way to the use of computers for personal productivity with the advent of the microprocessor. As the power of these microprocessors has grown exponentially, they have also been seen more recently as a vehicle for new media and socialization.

The ramifications of future computing technologies will be determined principally by two factors: the conception, development, and adoption of new applications that require significantly more computational power; and the ability of technology to address these demands. New applications are always difficult to anticipate, but it is less challenging to foresee the likely consequences for diffusion of this technology. Past experience with personal computers and telecommunications has shown that these technologies diffuse more rapidly in the developed world than in the developing world. It is difficult to foresee an increase in the political or ethical barriers to computing technologies beyond those seen today, and these are rapidly vanishing. On the remaining question of technology development, the odds-on favourite for the next 15 years remains traditional digital electronic computers based on semiconductor technology. Given the virtual certainty of continued progress in this area, it is hard to imagine a scenario in which a competing technology (quantum-switch-based computing, molecular computers, or something else) could offer a significant performance advantage at a competitive price. But the longer-term, traditionally elusive question in the period after 2015 is: How long will traditional silicon computing last? And when, if ever, will a competing technology become available and attractive?

If an alternative computing technology becomes sufficiently attractive, the economic effects of technology substitution on the current semiconductor industry and adjacent industries must be considered. For example, major industry players may be faced with a choice between cannibalizing their existing market opportunities in favour of these new, future technologies, competing head-on with new players, or simply acquiring them. Most important, given the very different architectural approaches of these technologies and the classes of problems for which they are best suited, what will be the effect on future applications? The promises of nanotechnology may indeed become a reality in the period after 2015, but it will face these competitive challenges before its significance becomes global.

INTEGRATED MICROSYSTEMS AND MEMS

MEMS is less an application area in itself than a manufacturing or fabrication technique that enables other application areas. Many authors use MEMS as shorthand to imply a number of particular application areas. As it is used here, MEMS is a "top-down" fabrication technology that is especially useful for integrating mechanical and electrical systems together on the same chip. It is grouped in the category of integrated microsystems because these same MEMS techniques can be extended in the future to also help integrate biological and chemical components on the same chip, as discussed below. Thus far, MEMS techniques have been used to make some functional commercial devices such as sensors and single-chip measurement devices. Many researchers have used MEMS technologies as analytical tools in other areas of nanotechnology such as the ones discussed here.

Smart Systems-on-a-Chip (and Integration of Optical and Electronic Components)

Simple electro-optical and chemical sensor components have already been successfully integrated onto logic and memory chip designs in research and development labs. Likewise, radio frequency component integration in wireless devices is already being produced in mass quantities. Some companies have products capable of doing elementary DNA testing.

The 1999 ITRS predicts the introduction of chemical sensor components with logic in commercial designs by 2002, with electro-optical component integration by 2004, and biological systems integration by 2006. Given these predictions, there is clearly time for relatively complex integrated systems and applications to develop within the 2015 timeline. These advances could enable many applications where increased integrated functionality can become ubiquitous as a result of lower costs and micro-packaging.

Micro/Nanoscale Instrumentation and Measurement Technology

Instrumentation and measurement technologies are some of the most promising areas for near-term advancements and enabling effects. As optical, fluidic, chemical, and biological components can be integrated with electronic logic and memory components on the same chip at marginal cost, drug discovery, genetics research, chemical assays, and chemical synthesis are all likely to be substantially affected by these advances by 2015.

Some of the first applications of nanoscale (and microscale) instruments were as basic sensors for acceleration (such as those used in airbags), pressure, etc. Small, microscale, special-purpose optical and chemical sensors have been used for some time in sophisticated laboratory equipment, along with microprocessors for signal processing and computation. Already, companies have produced products that allow for basic DNA analysis, and that can assist in drug discovery. As these sensors become more sophisticated and more integrated with computational capability (with the aid of systems-on-a-chip), their utility should grow tremendously, especially in the biomedical arena.

Broader Issues and Implications

There are several advantages of nanotechnology for integrated systems in general (and instrumentation and measurement systems as a subset of these). First, existing semiconductor technologies will likely allow the volume manufacture of integrated smart systems that can be produced at low enough cost to be considered disposable. Second, the massive parallelism afforded by this same technology allows for the rapid analysis (with integrated computation) of very complex samples (such as DNA), the processing of large numbers of samples, and the recognition of large numbers of agents (e.g., infectious agents and toxins). Devices with these properties are already of tremendous utility in the biomedical arena for drug testing, chemical assays, etc. In addition, they will likely find utility in a variety of industrial applications.

Integrated micro/nanosystems are already starting to affect applications where miniaturization of components, subsystems, and even complete systems is significantly reducing device size, power, and consumables while introducing new capabilities. This area lends itself naturally to the confluence of all the broad areas discussed in this report (biotechnology, materials, and nanotechnology).

The next five to ten years will likely see the integration of computational capabilities with biological, chemical, and optical components in systems-on-a-chip. At the same time, advances in biotechnology should drive applications for drug discovery and genomics, as well as the basic understanding of many other phenomena. Advances in biomaterials will likely produce biologically compatible packaging, capable of isolating substances from the body in a time-controlled fashion (e.g., for drug delivery). The confluence of these capabilities could allow for continued development of microscale and nanoscale systems that could continue to be introduced into the body to perform basic diagnostic functions in a minimally invasive way, providing new abilities to remedy health problems.

Other possible applications include: pervasive, self-moving sensor systems; nanoscrubbers and nanocatalysts; even inexpensive, networked "nanosatellites." For example, so-called "nanosatellites" are targeting order-of-magnitude reductions in both size and mass (e.g., down to 10 kg) by reducing major system components using integrated microsystems.

If successful, this could economize current missions and approaches (e.g., communication, remote sensing, global positioning, and scientific study) while enabling new missions (e.g., military tactical space support and logistics, distributed sparse aperture radar, and new scientific studies). In addition, advances could empower the proliferation of currently controlled processing capabilities (e.g., nuclear isotope separation) with associated threats to international security. Progress will likely depend on investment levels as well as continued S&T development and progress.

MOLECULAR MANUFACTURING AND NANOROBOTS

TECHNOLOGY

A number of experts have put forth the concept of molecular manufacturing where objects are assembled atom by atom (or molecule by molecule). Bottom-up molecular manufacturing differs from microtechnology and MEMS in that the latter employ top-down approaches using bulk materials using macroscopic fabrication techniques.

To realise molecular manufacturing, a number of technical accomplishments are necessary. First, suitable molecular building blocks must be found. These building blocks must be physically durable, chemically stable, easily manipulated, and (to a certain extent) functionally versatile. Several workers in the field have suggested the use of carbon-based diamond-like structures as building blocks for nano-mechanical devices, such as gears, pivots, and rotors.

Other molecules could also be used to build structures, and to provide other integrated capabilities, such as chemically reactive structures. Much additional work in the area of modeling and synthesis of appropriate molecular structures is needed, and a number of groups are working to this end. Dresselhaus and others are fabricating suitable molecular building blocks for these structures.

The second major area for development is in the ability to assemble complex structures based on a particular design. A number of researchers have been working on different approaches to this issue. Different techniques for physical placements are under development. One approach by Quate, MacDonald, and Eigler uses atomic-force or molecular microscopes with very small nanoprobes to move atoms or molecules around with the aid of physical or chemical forces. An alternative approach by Prentiss uses lasers to place molecules in a desired location. Chemical assembly techniques are being addressed by a number of groups, including Whiteside's approach to building structures one molecular layer at a time.

A third major area for development within molecular manufacturing is systems design and engineering. Extremely complex molecular systems at the macro scale will require substantial subsystem design, overall system design, and systems integration, much like complex manufactured systems of the present day. Although the design issues are likely to be largely separable at a subsystems level, the amount of computation required for design and validation is likely to be quite substantial. Performing checks on engineering constraints, such as defect tolerance, physical integrity, and chemical stability, will be required as well.

Some workers in the area have outlined a potential path

for the evolution of molecular manufacturing capability, which is broken down by overall size, type of fabrication technology, system complexity, component materials used, etc.

Some versions of this concept foresee the use of massively parallel nanorobots or scanning nanoprobes to assemble structures physically (with 100 to 10,000 molecular parts). Other, more advanced concepts incorporate chemical principles and use simple chemical feedstocks to achieve much larger devices on the order of 10^8 to 10^9molecular parts.

Ostensibly, as each of these techniques matures (or fails to develop), more systems and engineering-level work must be done before applications can be realized on a significant scale.

Although molecular manufacturing holds the promise of significant global changes (such as retraining large numbers of manufacturing workforces, opportunities for new regions to vie for dominance in a new manufacturing paradigm, or a shift to countries that do not have legacy manufacturing infrastructures), it remains the least concrete of the technologies discussed here.

Significant progress has been made, however, in the development of component technologies within the first regime of molecular manufacturing, where objects might be constructed from simple molecules and manufactured in a short amount of time via parallel atomic force microprobes or from simple self-assembled structures.

Although the building blocks for these systems currently exist only in isolation at the research stage, it is certainly reasonable to expect that an integrated capability could be developed over the next 15 years.

Such a system could be able to assemble structures with between 100 and 10,000 components and total dimensions of perhaps tens of microns.

A series of important breakthroughs could certainly cause progress in this area to develop much more rapidly, but it seems very unlikely that macro-scale objects could be constructed using molecular manufacturing within the 2015 timeframe.

BROADER ISSUES AND IMPLICATIONS

The present period in molecular manufacturing research is extremely exciting for a number of reasons. First, many workers have begun to experimentally demonstrate basic capabilities in each of the core areas outlined above. Second, continued progress and ongoing challenges in the area of top-down microelectronics manufacturing are pushing existing capabilities closer to the nanoscale regime. Third, the understanding of fundamental properties of structures at the nanoscale has been greatly enhanced by the ability to fabricate very small test objects, analyse them experimentally using new capabilities, and understand them more fundamentally with the aid of sophisticated computational models. At the same time, many visionaries have advanced notions about potential applications for molecular manufacturing. But because experimental capabilities are in their infancy (as many workers have pointed out), it is extremely difficult to foresee many outcomes, let alone assess their likelihood.

International competition for dominance or even capability in cutting-edge nanotechnology may still remain strong, but current investments and direction indicate that the United States and Europe may retain leadership in most of this field. Progress in nanotechnology will depend heavily on R&D investments; countries that continue to invest in nanotechnology today may lead the field in 2015. In 1997, annual global investments in nanotechnology were as follows: Japan at $120 million, the United States at $116 million, Western Europe at $128 million, and all other countries (former Soviet Union, China, Canada, Australia, Korea, Taiwan, and Singapore) at $70 million combined. Funding under the U.S. National Nanotechnology Initiative is proposed to increase to $270 million and $495 million in 2000 and 2001, respectively. This would not preclude other countries from acquiring capabilities in nanotechnology or in using these capabilities for narrow technological surprise or military means. Given the difficulty in foreseeing outcomes and estimating likelihoods, however, it is also difficult to extrapolate predictions of specific threats and risks from current trends.

Chapter 10

Scanning Tunneling Spectroscopy

Scanning tunneling spectroscopy (STS) studies the local electronic structure of a sample's surface. The electronic structure of an atom depends upon its atomic species (whether it is a gallium atom or an arsenic atom, for instance) and also upon its local chemical environment (how many neighbors it has, what kind of atoms they are, and the symmetry of their distribution). STS encompasses many methods: taking "topographic" (constant-current) images using different bias voltages and comparing them; taking current (constant-height) images at different heights; and ramping the bias voltage with the tip positioned over a feature of interest while recording the tunneling current. The last example results in current vs. voltage (I-V) curves characteristic of theelectronic structure at a specific x,y location on the sample surface. STMs can be set up to collect I-V curves at every point in a data set, providing a three-dimensional map of electronic structure. With a lock-in amplifier, dI/dV (conductivity) or dI/dz (work function) vs. V curves can be collected directly. All of these are ways of probing the local electronic structure of a surface using an STM.

Force vs. Distance Curves

Force vs. distance (F vs. d) curves are used to measure the vertical force that the tip applies to the surface while a contact-AFM image is being taken. This technique can also be used to analyse surface contaminants' viscosity, lubrication thickness, and local variations in the elastic properties of the surface.

Strictly speaking, a force vs. distance curve is a plot of the deflection of the cantilever versus the extension of the piezoelectric scanner, measured using a position-sensitive photodetector. The van der Waals force curve represents just one contribution to the cantilever deflection. Local variations in the form of the F vs. d curve indicate variations in the local elastic properties. Contaminants and lubricants affect the measurement, as does the thin layer of water that is often present when operating an AFM in air.

In the laboratory, force vs. distance curves are quite complex and specific to the given system under study. The discussion here refers to Figure and represents a gross simplification, where shapes, sizes, and distances should not be taken literally.

With this in mind, consider the simplest case of AFM in vacuum. At the left side of the curve, the scanner is fully retracted and the cantilever is undeflected since the tip is not touching the sample. As the scanner extends, the cantilever remains undeflected until it comes close enough to the sample surface for the tip to experience the attractive van der Waals force. The tip snaps into the surface. Equivalently, the cantilever suddenly bends slightly towards the surface.

As the scanner continues to extend, the cantilever deflects away from the surface, approximately linearly. After full extension, at the extreme right of the plot, the scanner begins to retract. The cantilever deflection retraces the same curve (in the absence of scanner hysteresis) as the scanner pulls the tip away from the surface.

In air, the retracting curve is often different because a monolayer or a few monolayers of water are present on many surfaces. This water layer exerts a capillary force that is very strong and attractive. As the scanner pulls away from the surface, the water holds the tip in contact with the surface, bending the cantilever strongly towards the surface. At some point, depending upon the thickness of the water layer, the scanner retracts enough that the tip springs free. This is known as the snap-back point. As the scanner continues to retract beyond the snap-back point, the cantilever remains undeflected as the scanner moves it away from the surface in free space.

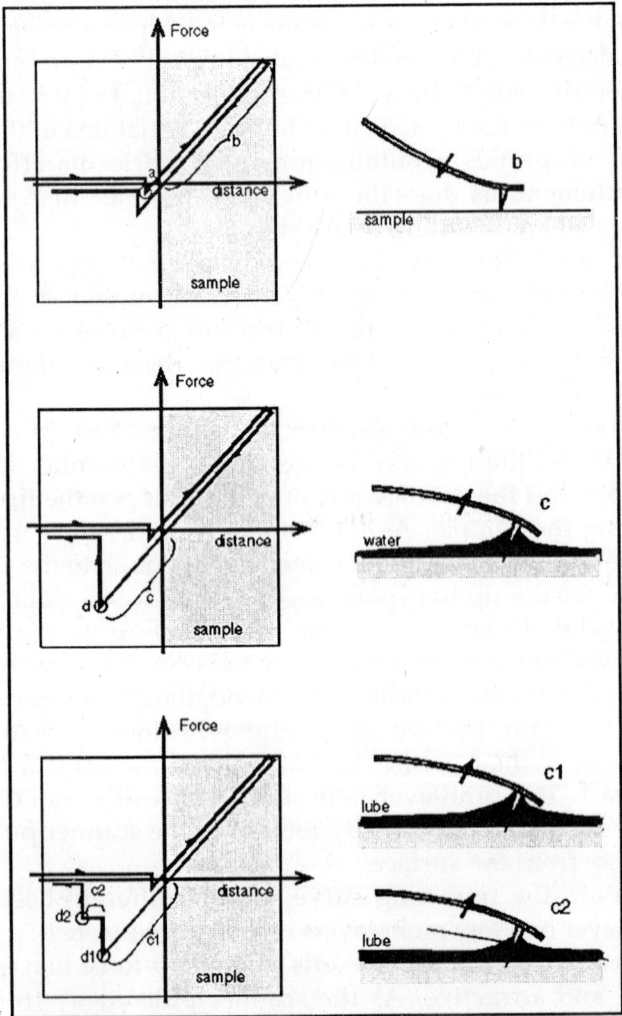

Fig. Forces vs. Distance Curves in Vacuum (top), Air (middle), and Air with a Contamination Layer (bottom)

If a lubrication layer is present along with the water layer, multiple snap-back points can occur, bottom. The positions and amplitudes of the snap-back points depend upon the viscosity and thickness of the layers present on the surface. Contact AFM can be operated anywhere along the linear portion of the force vs. distance curves, in regions (b) or (c). Operation in region (c) might be used for soft samples, to minimize the total force between the tip and the sample. Operating with the cantilever bent towards the surface is inherently a less stable situation, and maximum scan speeds may have to be reduced.

Note that operation in region (c) is still called contact mode, since the tip is touching the sample. Non-contact AFM is operated just to the left of point (a) on the F vs. d curve-just before the tip snaps into the surface.

In the linear region of the F vs. d curves (b), the slope is related to the elastic modulus of the system. When the cantilever is much softer than the sample surface, as is the case for non-destructive imaging, the slope of the curve mostly reflects the spring constant of the cantilever. However, when the cantilever is much stiffer than the sample surface, the slope of the F vs. d curve allows investigation of the elastic properties of the sample.

SPM ENVIRONMENTS

ULTRA-HIGH VACUUM

The first STMs were operated primarily in UHV to study atomically clean surfaces. Silicon has been the most extensively studied material, and obtaining images of the 7 x 7 surface reconstruction of Si (111) is often used as a benchmark for evaluating the performance of UHV STM. A major application of UHV STM is scanning tunneling spectroscopy. STS applied to atomically clean surfaces allows characterization of both topographic and electronic structure, without the added complication of contamination that is always present in air. Another application is the study of materials processes in situ-that is, in the same vacuum environment in which the materials are grown. In this way, processing may proceed without contaminating the sample.

Although the first scanning probe microscope was a UHV STM, the development of UHV AFM came much later.

Changing cantilevers and aligning the beam-bounce detection scheme - procedures that are not too difficult with an ambient AFM - were major obstacles to the development of a UHV AFM. However, the operation of both STM and AFM on the same vacuum flange is possible using piezoresistive cantilevers. A combined UHV STM/AFM enables scientists to work with conducting or non-conducting samples.

AMBIENT

The easiest, least expensive, and thus most popular environment for SPMs is ambient, or air. STM in air is difficult, since most surfaces develop a layer of oxides or other contaminants that interfere with the tunneling current. One class of materials for which ambient STM works well is layered compounds.

In graphite, MoS2, Nb3Se, and so forth, a clean, "fresh" surface can be prepared by peeling away older surfaces. AFM is indifferent to sample conductivity, and thus can image any solid surface in air, including the contaminants.

LIQUID

Liquid cells for the SPM allow operation with the tip and the sample fully submerged in liquid, providing the capability for imaging hydrated samples. Operation in liquid can also reduce the total force that the tip exerts on the sample, since the large capillary force is isotropic in liquid.

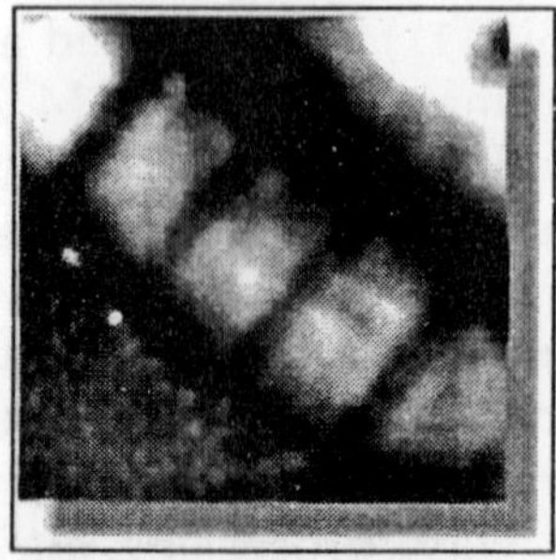

Fig. IC-AFM Image of Rabbit Muscle Myofibril taken in Liquid. Field of View 7 μm

A liquid environment is useful for a variety of SPM

applications, including studies of biology, geologic systems, corrosion, or any surface study where a solid-liquid interface is involved. AFM, LFM, and IC-AFM, FMM, and PDM images can be taken in a liquid environment. A IC-AFM image of rabbit muscle myofibril taken in liquid.

ELECTROCHEMICAL

Like UHV, electrochemical cells provide a controlled environment for SPM operation. Usually provided as an option for ambient SPMs, EC-SPMs consist of a cell, a potentiostat, and software. Applications of EC-SPM include real-space imaging of electronic and structural properties of electrodes, including changes induced by chemical and electrochemical processes, phase formation, adsorption, and corrosion as well as deposition of organic and biological molecules in electrolytic solution.

THE SCANNER

In virtually all scanning probe microscopes, a piezoelectric scanner is used as an extremely fine positioning stage to move the probe over the sample (or the sample under the probe). The SPM electronics drive the scanner in a type of raster pattern.

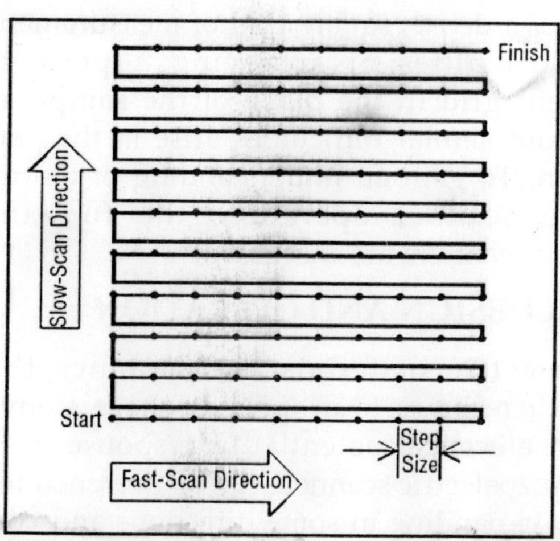

Fig. Scanner Motion During Data Acquisition

The scanner moves across the first line of the scan, and back. It then steps in the perpendicular direction to the second scan line, moves across it and back, then to the third line, and so forth. The path differs from a traditional raster pattern in that the alternating lines of data are not taken in opposite directions. SPM data are collected in only one direction - commonly called the fast-scan direction - to minimize line-to-line registration errors that result from scanner hysteresis. The perpendicular direction, in which the scanner steps from line to line, is called the slow-scan direction.

While the scanner is moving across a scan line, the image data are sampled digitally at equally spaced intervals. The data are the height of the scanner in z for constant-force mode (AFM) or constant-current mode (STM). For constant-height mode, the data are the cantilever deflection (AFM) or tunneling current (STM).

The spacing between the data points is called the step size. The step size is determined by the full scan size and the number of data points per line. In a typical SPM, scan sizes run from tens of angstroms to over 100 microns, and from 64 to 512 data points per line. (Some systems offer 1024 data points per line.) The number of lines in a data set usually equals the number of points per line. Thus, the ideal data set is comprised of a dense, square grid of measurements.

The difficulties of achieving a perfectly square measurement grid in the plane of the sample surface are explored. And similar difficulties arise in the perpendicular (z) direction. To gain an understanding of the mechanisms underlying scanner operation, the fundamentals of piezoelectric scanners are discussed.

SCANNER DESIGN AND OPERATION

Piezoelectric materials are ceramics that change dimensions in response to an applied voltage. Conversely, they develop an electrical potential in response to mechanical pressure. Piezoelectric scanners can be designed to move in x, y, and z by expanding in some directions and contracting in others. Piezoelectric scanners for SPMs are usually fabricated from lead zirconium titanate, or PZT, with various dopants added to create specific materials properties.

Scanners are made by pressing together a powder, then sintering the material. The result is a polycrystalline solid. Each of the crystals in a piezoelectric material has its own electric dipole moment. These dipole moments are the basis of the scanner's ability to move in response to an applied voltage.

After sintering, the dipole moments within the scanner are randomly aligned. If the dipole moments are not aligned, the scanner has almost no ability to move. A process called poling is used to align the dipole moments. During poling the scanners are heated to about 200°C to free the dipoles, and a DC voltage is applied to the scanner. Within a matter of hours most of the dipoles become aligned. At that point, the scanner is cooled to freeze the dipoles into their aligned state. The newly poled scanner can then respond to voltages by extending and contracting.

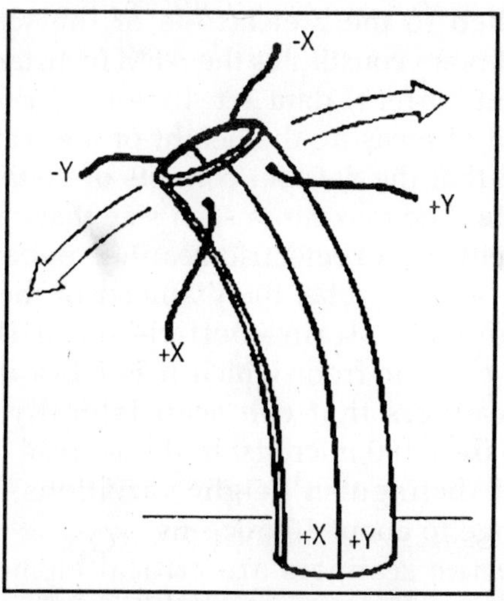

Fig. The Scanner Tube

Occasional use of the scanner will help maintain the scanner's polarization. The voltage applied to enact the scanning motion realigns the stray dipoles that relax into random orientation. If the scanner is not repoled by regular use, a significant fraction of the dipoles will begin to randomize (depolarize or depole) again over a period of weeks. Depoling is accelerated markedly if the scanner is subjected to temperatures above 150°C.

This means that if we want to add a heated stage to our SPM, we must isolate it thermally from the scanner. (The Curie temperature for PZT materials is about 150°C.) Many SPMs use variations of the simple tube design depicted in Figure. In this design, the scanner is a hollow tube.

Electrodes are attached to the outside of the tube, segmenting it electrically into vertical quarters, for +x, +y, -x, and -y travel. An electrode is also attached to the centre of the tube to provide motion in the z direction. When alternating voltages are applied to the +x and -x electrodes, for example, the induced strain of the tube causes it to bend back and forth in the x direction. Voltages applied to the z electrode cause the scanner to extend or contract vertically. In most cases, the voltage applied to the z electrode of the scanner at each measurement point constitutes the AFM (constant-force) or the STM (constant-current) data set. In some cases, an external sensor is used to measure the height of the scanner directly). It is assumed that the data set consists of voltages applied to the z electrode. The maximum scan size that can be achieved with a particular piezoelectric scanner depends upon the length of the scanner tube, the diameter of the tube, its wall thickness, and the strain coefficients of the particular piezoelectric ceramic from which it is fabricated. Typically, SPMs use scanners that can scan laterally from tens of angstroms to over 100 microns. In the vertical direction, SPM scanners can distinguish height variations from the sub-angstrom range to about 10 microns.

Piezoelectric scanners are critical elements in SPMs, valued for their sub-angstrom resolution, their compactness, and their high-speed response. However, along with these essential properties come some challenges.

SCANNER NONLINEARITIES

As a first approximation, the strain in a piezoelectric

scanner varies linearly with applied voltage. (Strain is the change in length divided by the original length) The following equation describes the ideal relationship between the strain and an applied electric field:

$$S = d\,E$$

where s is the strain in Å/m, E is the electric field in V/m, and d is the strain coefficient in Å/V. The strain coefficient is characteristic of a given piezoelectric material.

In practice, the behaviour of piezoelectric scanners is not so simple-the relationship between strain and electric field diverges from ideal linear behaviour.

INTRINSIC NONLINEARITY

Suppose the applied voltages start from zero and gradually increase to some finite value. If the extension of the piezoelectric material is plotted as a function of the applied voltage, the plot is not a straight line, but an s-shaped curve, as seen in Figure. The straight line in the graph is a linear fit to the data.

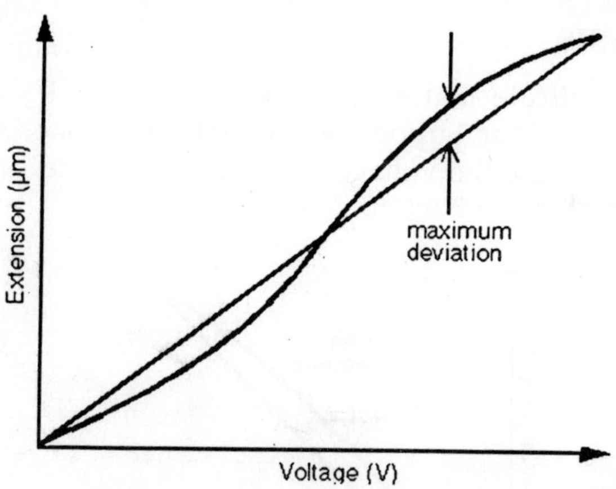

Fig. Intrinsic Nonlinearity of a Scanner

The intrinsic nonlinearity of a piezoelectric material is the ratio of the maximum deviation y from the linear behaviour to the ideal linear extension y at that voltage. In other words, the nonlinearity is the change in y divided by y, expressed as a percentage. Intrinsic nonlinearity typically ranges from 2 per cent to 25 per cent in the piezoelectric materials used in SPM systems. In the plane of the sample surface, the effect of intrinsic nonlinearity is distortion of the measurement grid. Because the scanner does not move linearly with applied voltage, the measurement points are not equally spaced. As a result, an SPM image of a surface with periodic structures will show non-uniform spacings and curvature of linear structures. Less regular surfaces may not show recognizable distortion, even though the distortion is present.

Perpendicular to the plane of the sample surface (in the z direction), intrinsic nonlinearity causes errors in height measurements. Height is usually calibrated for SPMs by scanning a sample with a known step height. By reading the voltage applied to the z electrode as it traverses the step, the z component of the strain coefficient can be calculated. However, if that strain coefficient is directly applied to measure a feature different in height from the calibrated height, the intrinsic nonlinearity of the scanner will generate errors.

HYSTERESIS

To complicate matters, piezoelectric ceramics display hysteretic behaviour. Suppose we start at zero applied voltage, gradually increase the voltage to some finite value, and then decrease the voltage back to zero.

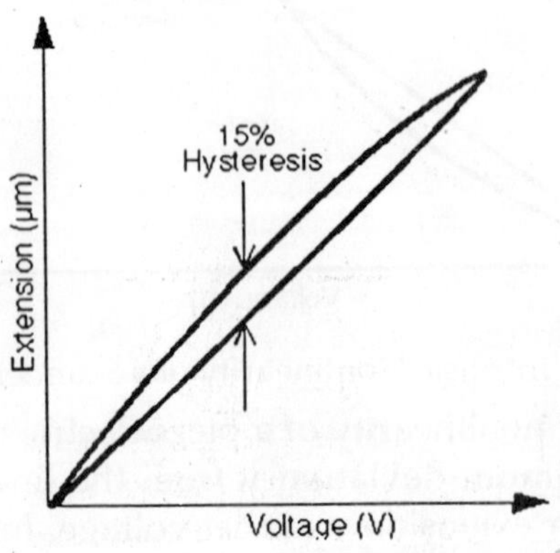

Fig. Hysteresis of a Scanner

If we plot the extension of the ceramic as a function of the applied voltage, the descending curve does not retrace the ascending curve - it follows a different path. It is assumed that the voltages change very slowly.

The hysteresis of a piezoelectric scanner is the ratio of the maximum divergence between the two curves to the maximum extension that a voltage can create in the scanner (the change in Y/Ymax). Hysteresis can be as high as 20 per cent in piezoelectric materials.

As mentioned SPM data are usually collected in one direction to minimize registration errors caused by scanner hysteresis. Data collected in the return direction would be shifted slightly, distorting the measurement grid. Since most SPMs today have provisions for taking scans in any fast-scan direction, we can observe hysteresis in the plane of the sample by comparing data sets taken in opposite fast-scan directions. For example, we could compare a data set taken left to right with a data set taken right to left. Scanner hysteresis would cause a shift between the data sets.

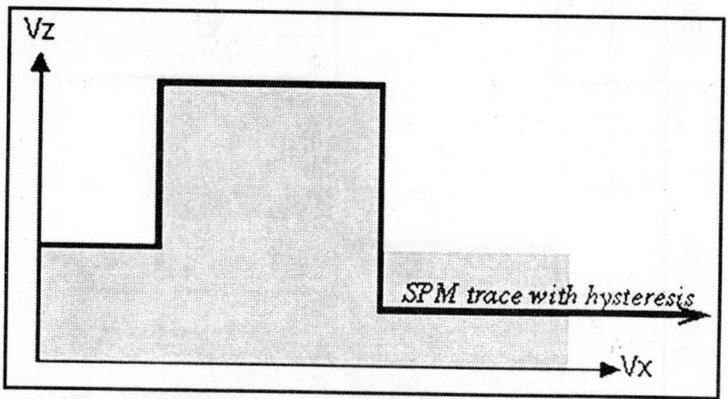

Fig. Effects of Hysteresis on a Step

Hysteresis in the direction perpendicular to the plane of the sample causes erroneous step-height profiles. Referring to the curve of Figure, we see that if the scanner is going up a step in the z direction, a certain voltage is required to allow the scanner to contract. But going down the same step the scanner extends, and extension takes more voltage than contraction for the same displacement. When the SPM image is represented by the voltage applied to the scanner. It is arbitrarily assumed that the strain coefficient in z was calibrated for a contracting scanner.

CREEP

When an abrupt change in voltage is applied, the piezoelectric material does not change dimensions all at once. Instead, the dimensional change occurs in two steps: the first step takes place in less than a millisecond, the second on a much longer time scale. The second step, delta xc, is known as creep.

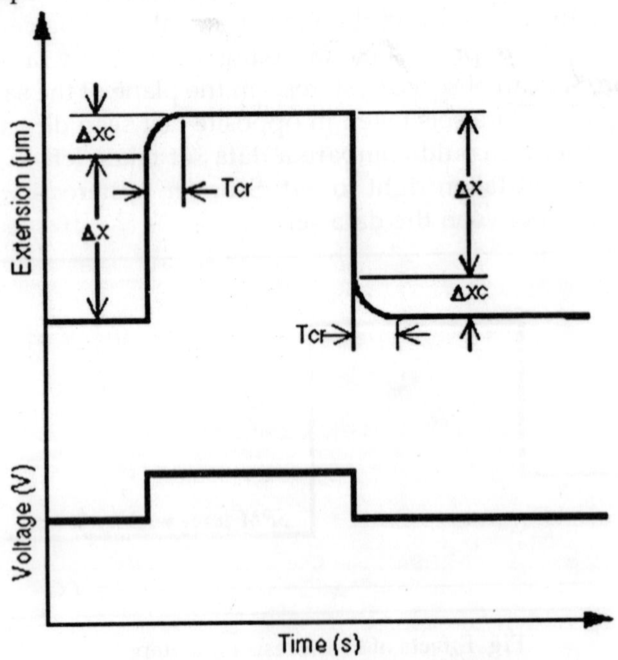

Fig. Plots Showing Creep of a Scanner.

Quantitatively, creep is the ratio of the second dimensional change to the first: the change in xc divided by the change in x. Creep is expressed as a percentage and is usually quoted with the characteristic time interval Tcr over which the creep occurs. Typical values of creep range from 1 per cent to 20 per cent, over times ranging from 10 to 100 seconds. The times involved during typical scans place the lateral motion of the scanner in the curved part of the response curve. As a result, two scans taken at different scan speeds show slightly different length scales (magnifications) when creep is present. We can trust only the measurement made at the scan speed that we used to calibrate our SPM when creep is present. Another effect of creep in the plane of the sample is that it can slow down our work when we try to zoom in on a feature of interest. Suppose we want to characterize a defect on the surface of our sample.

Most likely, we 'll first take a large scan to find the defect. Suppose we see the defect, very tiny, in a corner of the large scan. What we would prefer to do is take a much smaller scan - a high-resolution scan-centered about the defect. An SPM applies an offset voltage to move the scanner to its new centre position to take the second scan. Because we have applied a sudden voltage step, in this case an offset voltage in the plane of the sample, creep will cause we to miss our target. If we continue to collect scans without changing the offset voltage, the scanner will continue to relax until the defect eventually drifts (creeps) into the centre of our scan. The process will likely take a matter of minutes. To examine the effects of creep in the z direction. As the tip traverses the step from bottom to top, the scanner contracts immediately with a voltage corresponding to the full step height. However, over the next few seconds the scanner continues to contract slowly as creep occurs.

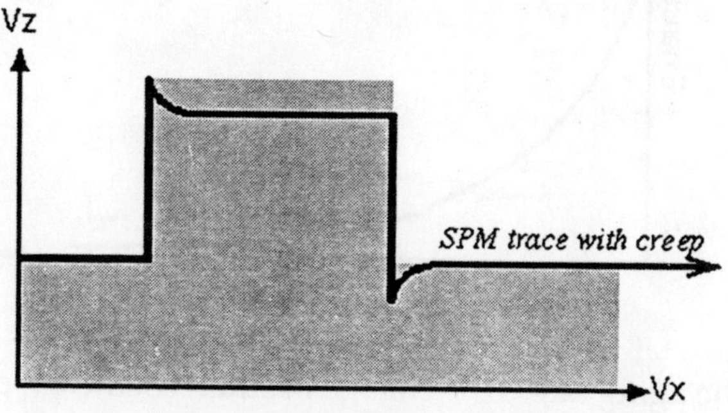

Fig. Effects of Creep on a Step.

To keep the tip in contact with the sample, an SPM has to apply a voltage in the other direction, counteracting the creep. When the tip traverses the step from top to bottom, the same process occurs. The scanner extends to accommodate the step, then continues to creep. An SPM must apply a voltage in the opposite direction to maintain contact between the tip and the sample.

The effect of creep on the profile of a step is a simple example. Consider a more typical sample with steep slopes that represent defects or grains or other surface features. Creep may cause an SPM image to look like it has ridges on one side of a feature and shadows on the other side of a feature. If there are shadows where the tip has just traveled down a steep slope, or ridges where the tip has just traveled up a steep slope, creep may be responsible. Reversing the fast-scan direction and taking the same image helps separate creep artifacts from true ridges and trenches.

AGING

The strain coefficient, d, of piezoelectric materials changes exponentially with time and use. The aging of a piezoelectric scanner, for cases of both high and low voltage usage.

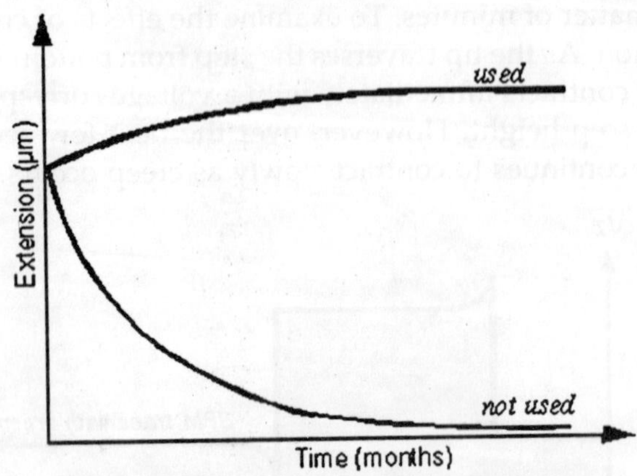

Fig. Aging of a Sscanner, with Use or Left Idle.

The aging rate is the change in strain coefficient per decade of time. When a scanner is not used, the deflection achieved for a given voltage gradually decreases. The aging rate of scanners for an SPM can produce a decrease in lateral strain coefficients (and therefore, an error in length measurements taken from SPM images) over time. When a scanner is used regularly, the deflection achieved for a given voltage increases slowly with use and time. These two phenomena are part of the same process. Each of the tiny crystals that compose the scanner has its own dipole moment. Repeated application of voltage in the same direction-such as the voltage applied during scanning-causes more and more of the dipoles to align themselves along the axis of the scanner. The amount of deflection achieved for a given voltage depends upon how many dipoles are aligned. Thus, the more the scanner is used, the farther the scanner will travel.

On the other hand, if the scanner is not used, the dipole moments of the crystals will gradually become randomly oriented again. As a result, fewer dipoles contribute to the deflection of the scanner. When we buy an SPM, the scanners have already been "poled," which means that they have already been exercised to the point where the deflection of the scanner is close to its maximum.

The dependence of the scanner deflection on time and usage means that the scanner may not be extending the same distance for a given applied voltage as it did when it was first calibrated. As a result, when we measure a feature on an SPM image, the values of lateral and vertical dimensions may be in error.

CROSS COUPLING

The term cross coupling refers to the tendency of x-axis or y-axis scanner movement to have a spurious z-axis component. It arises from several sources and is fairly complex. For example, the electric field is not uniform across the scanner. The strain fields are not simple constants, but actually complex tensors. Some "cross talk" occurs between x, y, and z electrodes. But the largest effect is geometric. Geometric cross coupling has its basis in the way piezoelectric scanners are constructed, usually as segmented tubes or as tripods.

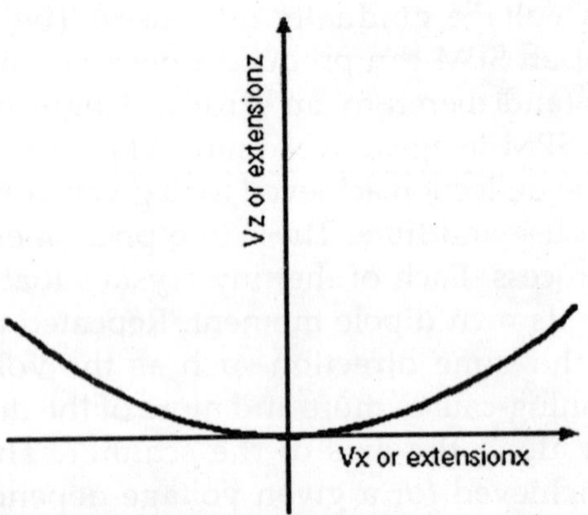

Fig. Cross Coupling Motion of the Scanner

The *x-y* motion of the scanner tube is produced when one side of the tube shrinks and the other side expands. As a result, a piezoelectric tube scans in an arc, not in a plane. A voltage applied to move the piezoelectric tube along the x or y axis (parallel to the surface of the sample) necessitates that the scanner extend and contract along the z axis (perpendicular to the surface of the sample) to keep the tip in contact with the sample.

A tripod scanner is designed with three mutually perpendicular bars or tubes glued together at one end. This design is also susceptible to cross coupling because the three bars of piezoelectric material are attached to one another. When the x bar extends or contracts, it causes rotation of the y and z bars. Cross coupling can cause an SPM to generate a bowl-shaped image of a flat sample. A profile of such an image is shown in Figure with an example of a step.

In interpreting, remember that the SPM image is based on the voltage required to compensate for the curvature generated by the arc of the scanner. The bowl shape may not always be evident in the final image because the curved background can be subtracted out using image-processing software. The best way to determine if the scanner is subject to cross coupling is to image a sample with a known radius of curvature, such as a lens. Software corrections can only flatten it or leave it alone. In the first case, we end up with a spuriously flat image, and in the second case the curvature of the scanner will be added to the curvature of the lens. The true curvature of a lens can be measured only when cross coupling is eliminated.

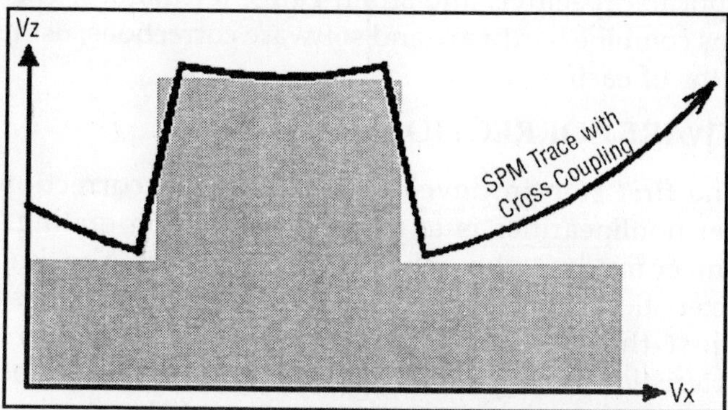

Fig. Effects of Cross Coupling on a Step

The example of a single step to demonstrate hysteresis, creep, and cross coupling in the vertical direction is used. To show a single image in the laboratory that illustrates each of these effects in isolation is virtually impossible. The sum of the effects of hysteresis, creep, and cross coupling in the image of a single step. (The aspect ratio of the tip may also contribute to the shape of the sidewalls)

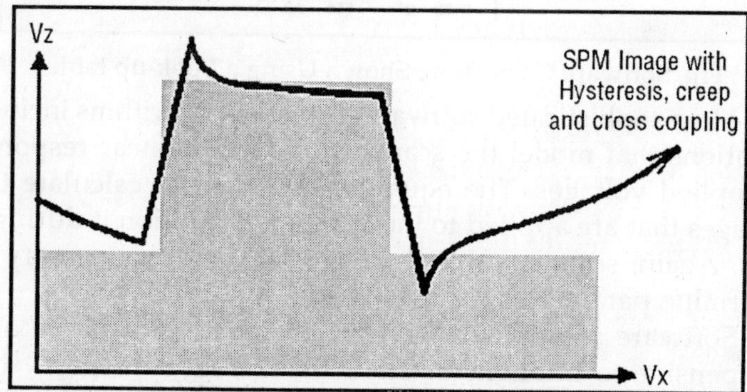

Fig. Effects of Hysteresis, Creep, and Cross Coupling on a Step

Traditionally, the nonlinear behaviour of piezoelectric scanners described above has been addressed imperfectly using software corrections. Some systems on the market use hardware solutions that eliminate most of the nonlinearities instead of correcting them. Hardware solutions are divided into optical, capacitive, and strain-gauge techniques. The best systems combine hardware and software corrections, using the strengths of each.

SOFTWARE CORRECTION

The first step in developing a software correction for scanner nonlinearities is to image a calibration grating. The system compares the measured data with the known characteristics of the grating. Then the system determines how to adjust the measured data to conform to the known characteristics. It stores that information in a file, or lookup table. Afterwards, the system can compensate for nonlinearity while collecting data by adjusting the voltage applied to the scanner in accordance with the file or the look-up table.

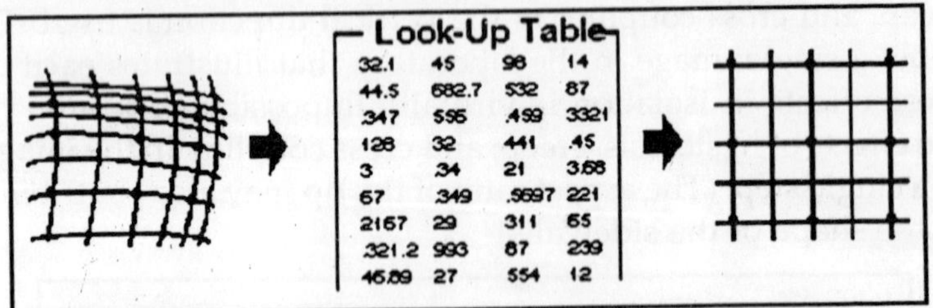

Fig. Software Correction, Shown Using a Look-up table

More sophisticated software correction algorithms include equations that model the scanner tube's non-linear response to applied voltages. The equations are used to calculate the voltages that are applied to the scanner to position it during a scan. Again, scanner calibration procedures are necessary to determine parameters for the algorithm.

Software solutions in general are relatively simple and inexpensive to implement. Their main disadvantage is that they compensate only partially for scanner nonlinearities. The corrections are strongly dependent upon the scan speed, scan direction, and whether the scanner was centered within its scan range during calibration. As a result, software corrections are most accurate only for scans that reproduce the conditions under which the calibration was done.

A scanner should be re-calibrated when scan conditions change. In practice, SPM systems that use only software corrections may vary from true linear behaviour by as much as 10 per cent. Often this deviation is unnoticed by the SPM user who calibrates and checks the calibration of the SPM with the same procedure every time.

HARDWARE CORRECTION

Systems with hardware solutions sense the scanner's actual position with external sensors. The signal read from the sensor of each axis is compared to a signal that represents the intended scanner position along that axis. A feedback system applies voltage to the scanner to drive it to the desired position. In this way, the scanner can be driven in a linear fashion. The external sensors themselves must be stable and immune to all sources of nonlinearity, since their purpose is to improve the linearity of the SPM.

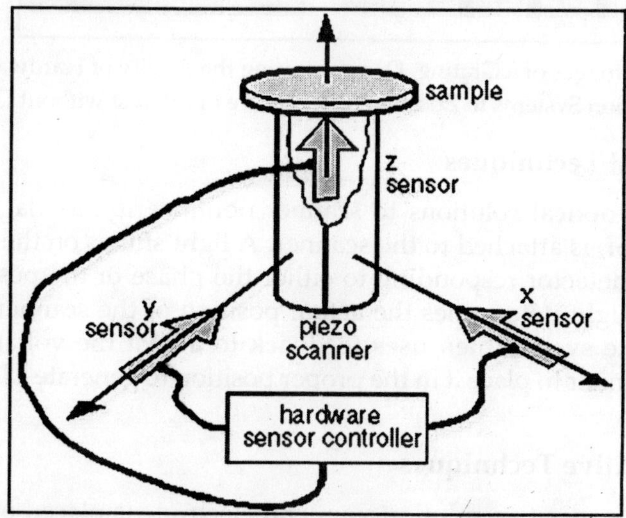

Fig. Schematic of a Scanner, Showing External Detectors

Because the scanner's actual position is measured, systems using hardware sensors compensate for intrinsic nonlinearity, hysteresis, creep, aging, and cross coupling. They can reduce the total nonlinearity of the system to less than 1 per cent.

Figure illustrates how SPMs equipped with hardware correction enable we to zoom in on a feature of interest in an image without the effects of creep.

The left image is a 40μm scan of a grating. The box in the image shows the intended location of the zoomed scan. The right image is the zoomed 2μm scan of that location. An image of the same area acquired at a later time would look identical to the right image, indicating that creep is not present in SPMs with hardware correction schemes.

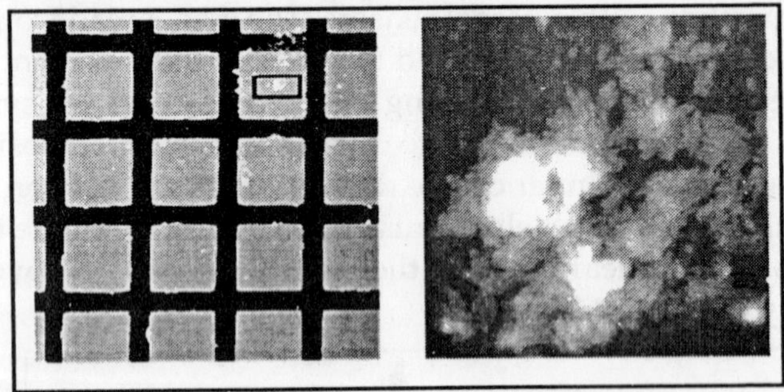

Fig. Images of a Grating, Demonstrating the Ability of Hardware Correction Systems to Zoom in on a Feature of Interest without Creep

Optical Techniques

In optical solutions to scanner nonlinearity, a "flag", or reflector, is attached to the scanner. A light shines on the flag, and a detector responding to either the phase or the position of the light determines the actual position of the scanner.

The system then uses feedback to adjust the voltage to the scanner to place it in the proper position to generate a linear scan.

Capacitive Techniques

In capacitive solutions, a metal electrode is placed on the scanner and another is mounted nearby. An electrical circuit then measures the capacitance between the electrodes, which varies with their separation. Having located the scanner in this way, the system uses feedback to adjust the scanner's position to compensate for nonlinearity.

Strain-Gauge Techniques

In strain-gauge solutions to scanner nonlinearities, a strain gauge is mounted on the scanner. The active element in a strain gauge is a piece of piezoresistive material-material whose resistance varies when it is under tensile or compressive stress. When the scanner moves, the resistance in the strain gauge changes.

The magnitude of the resistance of the strain gauge reflects the amount of bending in the scanner, thereby measuring its position. An SPM uses feedback to adjust the scanner's position and to generate a linear scan.

Tests for Scanner Linearity

If we suspect that nonlinear behaviour of the piezoelectric scanner has created artifacts in an SPM image, we can perform various tests. The symptoms and gives a test procedure to identify each source of nonlinearity. When we are evaluating an SPM, make sure that it has adequate compensation for intrinsic nonlinearity, hysteresis, creep, aging, and cross coupling. Otherwise, our SPM images may be distorted, the measurements we make from them may be erroneous, and the conclusions we draw from our measurements may lead we astray.

Intrinsic Nonlinearity

X-Y Plane

Symptom: Non-uniform spacings and curvature appear in structures that we know or believe to be linear.

Fig. Image of a Grating, Showing non-uniform Spacing and Curvature

Test: Image a calibration grating. The image should show consistent spacing and straight lines.

Cantilevers

Cantilevers and their tips are critical components of an atomic force microscope system because they determine the force applied to the sample and the ultimate lateral resolution of the system. Figure depicts a scanning electron microscope (SEM) image of a cantilever.

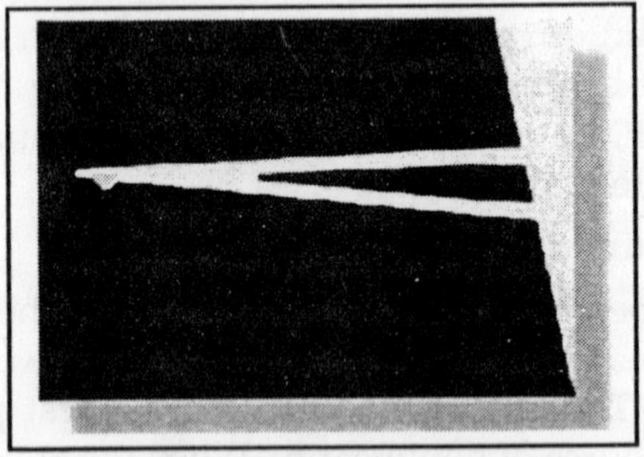

Fig. SEM image of a Cantilever

Integrated tip and cantilever assemblies can be fabricated from silicon or silicon nitride using photolithographic techniques.

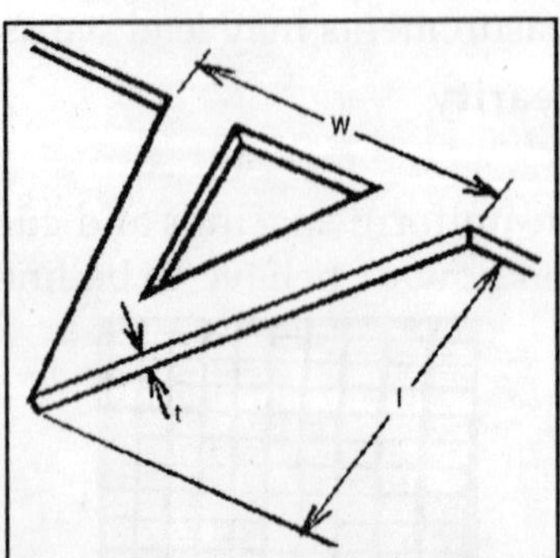

Fig. V-shaped cantilever, showing length (l), width (w), and thickness (t)

More than 1,000 tip and cantilever assemblies can be produced on a single silicon wafer. V-shaped cantilevers are the most popular, providing low mechanical resistance to vertical deflection, and high resistance to lateral torsion. Cantilevers typically range from 100 to 200μm in length, 10 to 40 μm in width, and 0.3 to 2 μm in thickness.

PROPERTIES OF CANTILEVERS

Atomic force microscopes require not only sharp tips, but also cantilevers with optimized spring constants - lower than the spring constants between atoms in a solid, which are on the order of 10 N/m. The spring constant of the cantilever depends on its shape, its dimensions, and the material from which it is fabricated. Thicker and shorter cantilevers tend to be stiffer and have higher resonant frequencies. The spring constants of commercially available cantilevers range over four orders of magnitude, from thousandths of a newton per meter to tens of newtons per meter. Resonant frequencies range from a few kilohertz to hundreds of kilohertz providing high-speed response and allowing for non-contact AFM operation.

HOW TO SELECT A CANTILEVER

The desirable properties for a cantilever depend on the imaging mode and the application. In contact mode, soft cantilevers are preferable because they deflect without deforming the surface of the sample. In non-contact mode, stiff cantilevers with high resonant frequencies give optimal results.

TIP SHAPE AND RESOLUTION

The lateral resolution of an AFM image is determined by two factors: the step size of the image, and the minimum radius of the tip. Consider an image taken with 512 by 512 data points. Such a scan 1 μm by 1μm would have a step size-and lateral resolution-of about 20Å (1 μm ÷ 512).

The sharpest tips available commercially can have a radius as small as 50Å. Because the interaction area between the tip and the sample is a fraction of the tip radius, these tips typically provide a lateral resolution of 10 to 20Å. Thus, the resolution of AFM images larger than 1 μm by 1μm is usually determined not by the tip but by the image's step size. The quoted *best resolution* of an AFM, however, depends upon how resolution is defined.

In the microscopy community, two asperities (peaks) are considered resolved if the image satisfies Rayleigh's criterion. In this application, Rayleigh's criterion requires that the height of the image dip at least 19 per cent between the asperities.

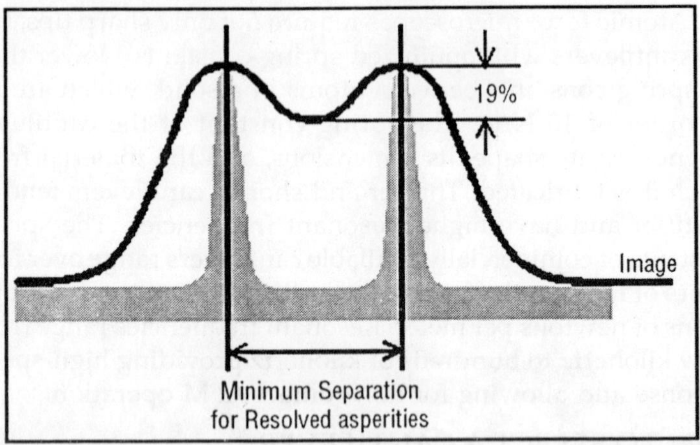

Fig. Definition of Lateral Resolution using Rayleigh's Criterion

To determine the lateral resolution of an SPM experimentally, the asperities are brought closer and closer together until the image no longer dips by 19 per cent between peaks. The minimum separation between resolved asperities determines the best lateral resolution of the system. Using this definition, the lateral resolution of an AFM with the sharpest tips commercially available is 10 to 20Å.

At first glance the figure of 10 to 20Å resolution seems to conflict with the ubiquitous images of atomic lattices in AFM brochures. The distinction between *imaging atomic-scale features* with accurate lattice spacing and symmetry, and *true atomic resolution* bears some comment.

An STM gives true atomic resolution. Because the dependence of the tunneling current on the tip-to-sample separation is exponential, only the closest atom on a good STM tip interacts with the closest atom on the sample, as shown in Figure, top. For AFMs, the dependence of cantilever deflection on tip-to-sample separation is weaker. The result is that several atoms on the tip interact simultaneously with several atoms on the sample.

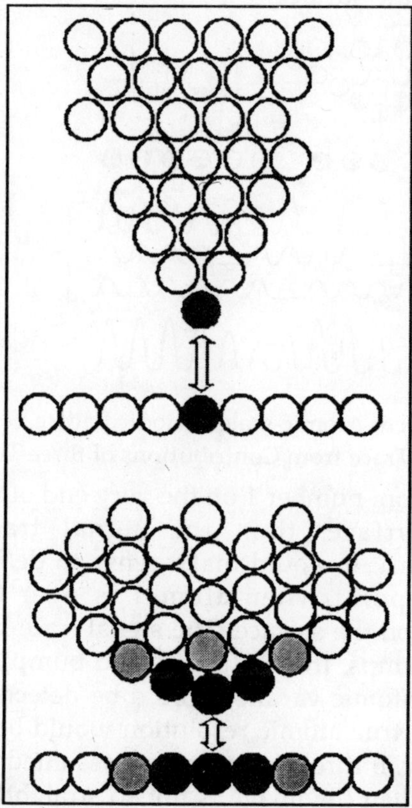

Fig. Interatomic Interaction for STM (top) and AFM (bottom). Shading shows Interaction Strength

Figure illustrates how this multiple interaction affects AFM images: AFMs cannot achieve the true atomic resolution needed to detect, for example, the atomic vacancy depicted in the figure.

The atomic vacancy is shifted in position and not well resolved in the sum signal trace. In the figure, three tip atoms, numbered 1 through 3, are assumed to interact with the atoms of the surface. Three signal traces are also shown, representing the contribution from each atom. The traces show signals varying as a function of time, where the signal reflects the interatomic force between the given tip atom and the atom beneath it on the surface.

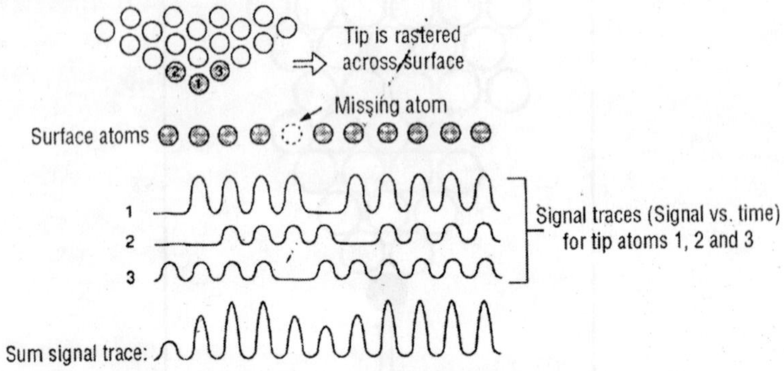

Fig. AFM Scan of Atomic-scale Periodic Lattice, Showing the Sum Signal Trace from Contributions of three Tip Atoms

If only atom number 1 on the very end of the tip interacts with the surface, then the signal trace would be straightforward: It would make sense to define time equals zero at the point when atom 1 is over the first atom encountered on the surface. The signal trace labeled 1 would show four bumps, then a hole (or no bump), then six more bumps. The atomic vacancy would be detected in its correct position, and true atomic resolution would be achieved.

However, if three tip atoms are assumed to interact with the surface, then the image is not so straightforward. In this case, it makes sense to define time equals zero when atom 3 is over the first atom encountered on the surface. When atom 3 encounters the first surface atom, some signal is registered, apparent from the signal trace of atom 3.

At the next instant in time (assuming a step size on the order of one atomic spacing), atom 1 detects the first surface atom, atom 3 detects the second surface atom, and atom 2 still registers no signal. At the third instant in time, all three tip atoms register signals which vary in intensity according to their distance from the surface.

This example shows that, since the tip atoms are displaced laterally, they produce signal traces that are offset from one and other. The vacancy appears in a different location for each signal trace. An image is the result of the sum of the contributions from all interacting tip atoms. Because the three interacting tip atoms are displaced from one and other laterally, there is no clear vacancy apparent in the sum signal. A minimum is apparent, but its location is shifted.

In this simple case, only three tip atoms were assumed to interact with the surface. In reality, the number of interacting tip atoms for contact AFM is large, and at each instant in time their interaction is with several surface atoms. For larger numbers of interacting atoms, features such as vacancies are even less well-defined and more shifted. Thus, while the periodicity of the lattice is reproduced, true atomic resolution is not achieved.

For both contact and non-contact imaging, we should choose a tip that is sharper than the smallest features on our sample. If the tip is larger than the surface features, an artifact know as *tip imaging* occurs. We may not always want to choose the sharpest tip available, since the sharpest tips are more expensive and can be less durable. We need the sharpest tips only when we require the best resolution.

For LFM, blunter tips can be preferable precisely because they present a larger interaction area between tip and sample. A larger area of interaction can produce a stronger lateral deflection of the cantilever. The larger cantilever deflection must be balanced against the lower lateral resolution delivered by the blunter tip.

AFM manufacturers offer microfabricated tips in three geometries: pyramidal, tetrahedral, and conical. Conical tips can be made sharp, with high aspect ratios (the ratio of tip length to tip width). Tip radii as small as 50Å have been observed. Pyramidal tips have lower aspect ratios and nominal tip radii of a few hundred angstroms, but they are more durable.

AFM tips are fabricated from silicon or silicon nitride. The fabrication process is different for the two materials. The features of tips made from each material are governed by the fabrication process as well as the material properties.

Silicon conical tips are made by etching into the silicon around a silicon dioxide cap. The high aspect ratio of conical tips makes them suitable for imaging deep, narrow features such as trenches, but they may break more easily than the pyramidal or tetrahedral geometries, as described above. Silicon also has the advantage that it can be doped, so that tips can be made electrically conducting.

Conducting tips are useful for controlling the bias between the tip and the sample, or for preventing unwanted charge buildup on the tip.

Silicon nitride tips are fabricated by depositing a layer of silicon nitride over an etched pit in a crystalline silicon surface. This method produces the pyramidal or tetrahedral tip geometry. The aspect ratio of a silicon nitride tip is thus limited by the crystallographic structure of the etch pit material, silicon. The tips are broader than conical silicon tips, making them sturdier but less suitable for imaging deep, narrow features. Silicon nitride is a harder material than silicon, which also makes silicon nitride tips more durable than silicon tips.

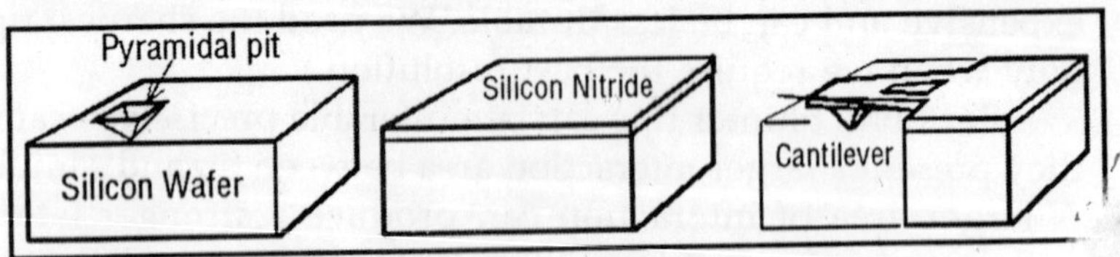

Fig. Fabrication of Silicon Nitride Tip

Silicon nitride films, however, contain residual stresses which make them deform as the film thickness increases. For this reason, applications that call for thick cantilevers - or cantilevers with high resonant frequencies - use silicon cantilevers. The thickness of silicon nitride cantilevers is usually less than one micron, whereas silicon cantilevers can be several microns thick.

A more specialized type of cantilever that has a tip grown in a scanning electron microscope is also available commercially. (A specialized tip can also be produced by "milling" an existing tip using a focused ion beam.) The SEM is used in an unusual way: to deposit contamination in a column. The tips are not microfabricated in bulk; they are grown one at a time on top of a bare cantilever or square-pyramidal tip. These tips have the advantage of almost unlimited aspect ratio. However, they are not normally sharp, and can bend or break easily.

IMAGE ARTIFACTS

SPM images are among the easiest to interpret of images generated by any microscopy technique. With an electron or optical microscope, contrast is based on complex electromagnetic diffraction effects. Determining whether a feature is protruding from the surface or recessed into it can be difficult with an image from an optical or electron microscope. SPMs, however, collect three-dimensional data. In an SPM image, a peak is unambiguously a peak, and a valley is clearly a valley. With an optical or electron microscope, artificial contrast can occur when a sample consists of reflecting material embedded in an absorbing matrix, for example. SPMs, on the other hand, are largely indifferent to variations in optical or electronic properties, and measure true surface topography.

Despite the apparent simplicity of images from SPMs, SPM images are subject to some artifacts of their own. Fortunately, artifacts are relatively easy to identify.

TIP CONVOLUTION

Most imaging artifacts in an SPM image arise from a phenomenon known as tip convolution or tip imaging. Every data point in an image represents a spatial convolution (in the general sense, not in the sense of Fourier analysis) of the shape of the tip and the shape of the feature imaged. As long as the tip is much sharper than the feature, the true edge profile of the feature is represented.

However, when the feature is sharper than the tip, the image will be dominated by the shape of the tip. Figure demonstrates the origin of tip convolution.

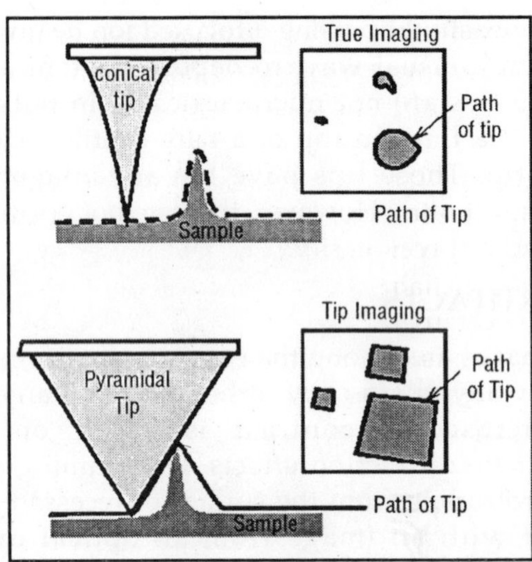

Fig. Comparison between True Imaging and Tip Imaging

The first commercial AFM tips were square pyramidal, formed by CVD deposition of Si3N4 on an etch pit in Si (100). The etch pit is bounded by (111) faces, which means that the resulting tip has a sidewall angle of about 62.5° along the flat side of the pyramid (45° along the corner edge of the pyramid).

If such a tip is used, the edge profiles of SPM images of all features with sides steeper than 62.5° are dominated by the profile of the tip. Tips with higher aspect ratios have been marketed successfully since then, but microfabricated tips remain limited to a sidewall angle of about 80°. Further enhancement of the sidewall angle of the tip can result in degraded durability, or even bending of the tip during scanning.

Because many samples have features with steep sides, tip imaging is a common occurrence in images. Sidewall angles on images should be measured routinely, to determine whether the slope is limited by that of the tip, or truly represents the topography of the sample. One consolation is that the height of the feature is reproduced accurately as long as the tip touches bottom between features. Thus, height measurements and roughness statistics remain fairly accurate. The lateral dimensions from a tip-imaged scan, on the other hand, can provide the user with only a maximum value. In other words, if we measure a tip-imaged feature as having a width of 200Å, we 'll know that the feature is at most 200Å wide.

To recognize tip imaging, look for a particular shape that is repeated throughout an image. The shape can be different sizes as the tip is convolved with features of different sizes, but it will always maintain the same orientation. If we suspect tip imaging is occurring, we can rotate our sample and image it again. If the tip is dominating the image, the orientation of the tip shape will be the same before and after rotation.

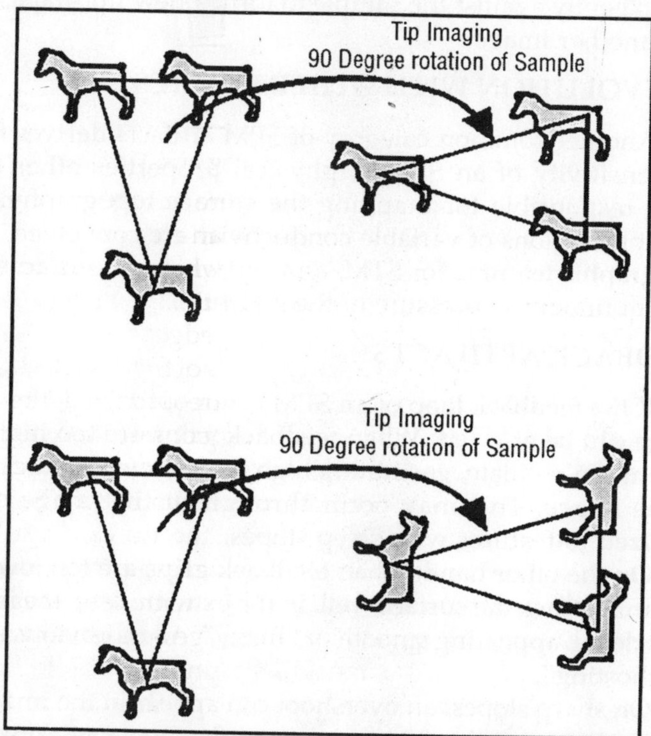

Fig. Top down Image showing Characteristically Shaped Features

If the image is a true representation of the surface, the shapes in the image will rotate along with the sample. This test is represented schematically in Figure. If our image is dominated by tip-convolution effects, change to another tip. For STM, the imaging portion of the tip is formed by an atom or cluster of atoms at the end of a long wire. Because the dependence of the tunneling current upon the tip-to-sample distance is exponential, the closest atom on the tip will image the closest atom on the sample.

If two atoms on the tip are equidistant from the surface, all of the features in the image will appear doubled. This is an example of multiple-tip imaging. The best way to alleviate this problem is to apply a voltage pulse to change the tip configuration by field emission. Alternatively, we can press the tip gently against the sample to form a new tip shape, and take another image.

CONVOLUTION WITH OTHER PHYSICS

Another common category of SPM artifacts derives from the sensitivity of an SPM to physical properties other than those responsible for mapping the surface topography. For example, regions of variable conductivity are convolved with topographic features for STM, and soft or elastic surfaces can deform under the pressure of the AFM tip.

FEEDBACK ARTIFACTS

If the feedback loop of an SPM is not optimized, the SPM image can be affected. When feedback gains are too high the system can oscillate, generating high frequency periodic noise in the image. This may occur throughout the image or be localized to features with steep slopes.

On the other hand, when feedback gains are too low, the tip cannot track the surface well. In the extreme case, the image loses detail, appearing smooth or "fuzzy". A less obvious effect is "ghosting".

On sharp slopes, an overshoot can appear in the image as the tip travels up the slope, and an undershoot can appear as the tip travels down the slope. This feedback artifact commonly appears on steep features, represented as bright ridges on the uphill side and/or dark shadows on the downhill side of the feature.

IMAGE PROCESSING CAPABILITIES

Sophisticated image-processing software is available on all commercial SPMs. Beautiful images can be created by introducing artificial light sources, 3-dimensional rendering, colour tables, and curvature enhancement algorithms, by retouching areas of bad data, by filtering environmental noise, and by magnifying or reducing the vertical scale of the image to enhance or minimize surface features of interest. We can flatten an image of a surface of a tiny sphere to look closely at the fine-scale roughness. These are some of the benefits of image-processing software.

Fig. 3-D Rendered Image of a Semiconductor Surface

However, when image processing and image enhancement are used carelessly or irresponsibly, data can be misrepresented.

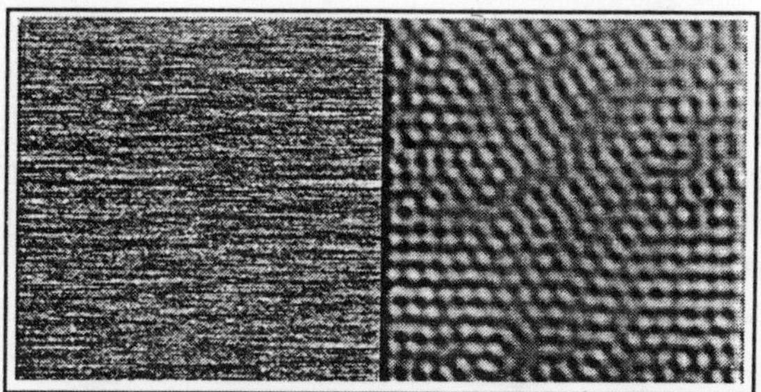

Fig. AFM Image of Random noise. Left Image is Raw Data

The best way to compare the quality of data sets from different SPMs is to compare unprocessed "raw" data in a gray-scale image. As an example of misuse of image processing, Figure shows a wholly specious image of a monolayer of an organic film, created entirely from random noise by a carefully crafted filter. Right image is a false image, produced by applying a narrow band pass filter to the raw data

KEY FEATURES OF SPMS

Considerations outside of the scientific realm are likely to be a part of our choice of an SPM. An instrument that is easy to use will have lower operating costs because a highly skilled operator is not required. Minimal training time is desirable in a multi-user facility. Finally, an instrument that is easy to run will be used more-and produce more results.

USER INTERFACE

Evaluate the complexity of the user interface. It should be laid out in a straightforward, manner with easy-to-use controls. For example, the user interface should have buttons with names that represent commonly performed functions such as "APPROACH" and "IMAGE". After we have seen someone else demonstrate the instrument, ask to operate it ourself. We are then in the best position to evaluate the simplicity of the user interface.

Some SPMs are built upon operating systems that allow multi-tasking. If we can process data while collecting additional data, our throughput will be increased measurably.

OPTICAL MICROSCOPE

All commercial SPMs now include optical microscopes to help monitor the tip-to-sample approach and to select the areas of interest on the sample surface.

A good optical microscope speeds up our work because it enables we to position the tip quickly and accurately, exactly where we want to take an SPM image.

If we intend to image very rough or oddly shaped samples (for example geological samples, or teeth), or cross sections of any kind that require landing the tip on a narrow edge, a good optical microscope is indispensable. Figure shows a real-time video image of a cantilever positioned over an integrated circuit.

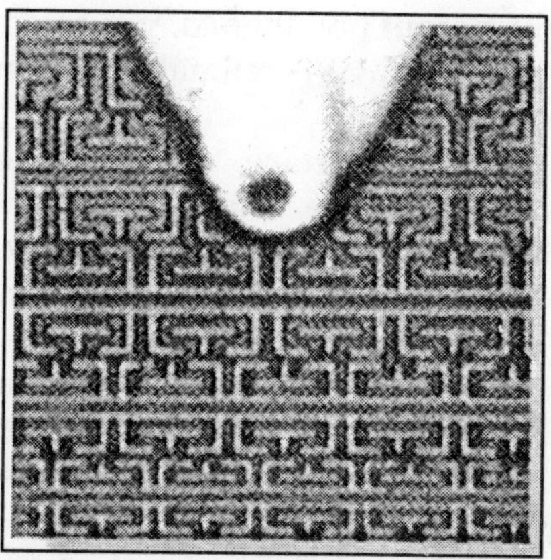

Fig. Real-time Video Image of a Cantilever Positioned Over an Integrated Circuit

When evaluating the optical microscope, look for the following features:

- Optical image clarity
- Useful magnification range
- Zoom lens
- Integrated video camera

Motorized focus and zoom controls are timesavers over manual models, but optical image quality is the most important. The magnification should enable we to navigate rough surfaces and cross sections, to a high magnification, where the field of view allows we to place the tip on micron-sized features. Note that a quoted magnification range is subject to "cheating". If we attach a larger monitor to the same video camera, we end up with higher magnification, but no better resolution. The best way to specify the optical microscope is by field of view or resolution.

PROBE HANDLING

Under normal operating conditions, an AFM tip should last for a couple of days, so changing the probe should be a straightforward process. However, the chips upon which cantilevers are mounted are very small and can be unwieldy to handle.

New designs permit pre-aligned, pre-mounted probes to be changed very quickly, with minimal alignment of the beam-bounce detection system.

As with the user interface, we should evaluate how difficult it is to change a probe. Change a probe ourself, instead of watching a highly skilled operator change it.

SYSTEM ACCESSIBILITY

If we intend to perform unique experiments with an SPM, we may be very interested in having access to the software, the electronics, and the mechanics of the system in order to make our own modifications.

Investigate the design of the system to see whether it is open enough for our customized work. Consider also the technical strength and availability of the technical support staff at the factory. We may want their advice more often than a user who is happy to operate the machine off-the-shelf.

Index